TECHNOLOGY AND HUMAN VALUES
Collision and Solution

Bruce O. Watkins
Professor of Electrical Engineering
Utah State University
Logan, Utah

Roy Meador
Science Writer
Ann Arbor, Michigan

ANN ARBOR SCIENCE
PUBLISHERS INC
P.O. BOX 1425 • ANN ARBOR, MICH. 48106

Softcover Edition Published 1978

230 Collingwood, P. O. Box 1425, Ann Arbor, Michigan 48106

Library of Congress Catalog Card No. 76-050985
ISBN 0-250-40241-6

Manufactured in the United States of America

Preface

On July 20, 1969 and July 20, 1976, two out-of-this-world events dramatically proved the brilliance of modern technology in performing feats that stagger the imagination. On the first date, technology arranged for two humans to walk on the surface of the moon and for millions on earth to watch them. The event on the later date earned an exceptionally congratulatory lead editorial in *The New York Times* under the title, "The Viking Miracle." The editorial praised the extraordinary technology that had controlled the Viking 1 rocket across more than 400 million miles of space to a remarkably efficient and safe landing on another planet. The *Times* tribute concluded with these words:

> Time as well as space have in a sense been conjoined. The instrument that has performed this miracle is that inhuman little robot, sending out its signals, a mere creature of another miracle whose power is even more spectacular than that of all the Vikings and the Mariners put together—the miracle of the human mind.

An advertisement in the same issue of the newspaper carried a headline commencing with this eloquent understatement: "Landing on Mars united a great many technologies." Such phenomenal accomplishments through combined technologies please the heart, inspire the mind, and even encourage the whispering of hope that technology will somehow succeed on this world in coping with the great challenges confronting mankind.

This book examines technology in its current relationship to such challenges and to the special human values espoused by that sometime worker of miracles, the human mind. What are the problems confronting mankind today? They are analyzed from human hunger to the spectre of nuclear destruction. The technological prospects for solving these problems are dissected. What technology can do is discussed as well as what it cannot do, and throughout it is related to the underlying human values that do not place a robot ambassador on Mars, but decide why and when it should be attempted.

The authors are a professor of engineering and a technical writer who independently reached the conclusion that scientific and nonscientific intellectuals should cease glaring at one another suspiciously across no man's land. The need for a bridge from science to the rest of the human community was starkly emphasized when one of the authors was shaken by an English professor's scornful dismissal of science. The need has been equally conspicuous in recent years for those in the technological professions to be alert to the opposite danger of pedantically dismissing ideas that cannot be scientifically analyzed and explained.

Nonscientists as well as scientists and engineers should benefit from this frank look at the state of technology today. Among other dividends, they might discover that technology is not incompatible with abstract ideas or devoid of close, working relationships with human values.

In a civilization of growing complexity, it becomes important for professionals in every discipline to acknowledge that they share common human values. Without input from all groups, the resulting values turn out as half-baked loaves.

The authors believe that science in particular must be leavened by humanistic concerns since science unaccompanied by human values may produce monsters. Human values are mankind's center of gravity. Their moral force is indispensable to protect us from the consequences E. B. White warned about in his short story, "Time Past, Time Future." Two American officers, Trett and Obblington, were on a space platform with a supply of weapons. Their assignment: guard America, keep the peace. But outside the pull of earth's gravity, the officers found they also didn't feel the pull of conscience or the pull of duty. So eventually, for fun, they used their weapons to destroy the earth.

In White's morality tale, the earth came to an end through a failure of human values. To keep truth from being more sinister than fiction, we think that technologists must see to it that they and their colleagues do not occupy platforms outside the pull of human values.

To assure that technologists are on the same wave lengths as those values, the authors examine the thesis that technologists (used broadly in this book to include both scientists and engineers) should make a special point of learning all they can—that they should explore history, art, philosophy, literature, and ideas, as well as technology. The authors have sought to show how values derived from other disciplines or from human nature can be not only beneficial but prerequisites for scientists attempting responsible management of their careers. These lines from a poem, "Footnotes for a Centennial," by Christopher Morley, provide a starting place:

> Science gives old riddles a new name
> But underneath, and still the same,
> Finds everything converging into one;
> ...whoever has humbly known
> Some worship of his own
> Has honor for all creeds.

The corollary is not neglected that all creeds should learn about technology sufficiently to honor it as well. Humanists, in other words, cannot afford in a technological civilization to remain ignorant about science.

Ultimately, intellectual progress and learning seem to take deliberate aim at the goals suggested by Morley: human thought converging and honor for all creeds. Technologists and scholars are invited to join the endeavor to converge. They have nothing to lose but their trenches, and no man's land to gain.

The authors are grateful to colleagues and friends who assisted with their counsel and encouragement. Elizabeth Hecht, with her tireless and dedicated efforts at the typewriter, earned and receives our special thanks.

Bruce Watkins
Logan, Utah

Roy Meador
Ann Arbor, Michigan

Table of Contents

"What has gone wrong with the West? A good many sensible men have thought about it for a long time now, and a good many of their answers overlap. Probably they are giving us some part of the truth. As usual, diagnosis is much easier than doing anything about the condition: But we shall not stand much chance unless we get the diagnosis right.

...it does no harm to be on the look-out for a change, a new and perhaps startling one, in our mental universe. Reason hasn't stopped. It may be on the point of surprising us. The encouraging thing for us western characters is that this is more likely to happen in plural societies than in more homogeneous ones. It is more likely to happen here."

C. P. Snow
"Grounds for Hope?"
Address at New York University
1976

1. Schism Between the Cultures

"At present we are making do in our half-educated fashion, struggling to hear messages, obviously of great importance, as though listening to a foreign language in which one only knows a few words."

C. P. Snow
The Two Cultures

THE NECESSITY OF KNOWING

There is irony in the fact that during the age of communication satellites and international mass media, one of mankind's most nagging problems is communication.

Technologists (scientists and engineers) have found it especially difficult to communicate with those not trained in mathematics and the special languages of technology. Many technologists have simply stopped trying to cross the gulf between themselves and others—either to teach or learn. They find it too difficult. Albert Einstein addressed the problem more than forty years ago, stressing his regret that "the activity of the individual investigator should be confined to a smaller and smaller section of human knowledge." "Every serious scientific worker is painfully conscious of this involuntary relegation to an ever-narrowing sphere of knowledge," wrote Einstein, adding that "we have all suffered under this evil, without making any effort to mitigate it" (1).

Improvement seems as distant as ever, but the danger of non-communication among men and silence among professions is worse than ever. Technology has become increasingly refined and intricate. But in the process of multiplying efficiency, specialists in one field have found it harder and harder to communicate with specialists in unrelated fields, and hardest of all to make sense when communicating with nontechnological intellectuals and the general public.

The problem of communication is aggravated by the fact that the demands of specialization often leave no energy or interest for the intellectual fertilization derived from exploring other fields and cultures.

"There are no chaste minds. Minds copulate wherever they meet," wrote philosopher Eric Hoffer (2). Hoffer was indulging in wishful thinking. Suspicion is strong that many professional minds meet, bow politely, and hurry on without so much as an exploratory tickle. After a hard day's science, the mind is less disposed to flirt with new horizons than to remove its shoes, have a martini, and take a nap before dinner.

Specialization, particularly in technical matters, seems firmly established as the wave of the present and the flood of the future. To keep all systems functioning, specialists rather than generalists are called for in constantly growing numbers. Physicist Edward Teller was praising, not deriding, when he described the specialist as "someone who knows more and more about less and less until he knows everything about nothing." But such deliberately concentrated and limited "knowing" creates severe communications obstacles, with specialists talking mainly to themselves.

Technologists aren't the only ones who specialize. Philosophers, artists, historians, oystermen, drama critics—most contemporary professionals—tend to concentrate more and more on less and less. So much so, omphaloskepsis (contemplation of the navel) seems the overwhelming, inward-turning, modern impulse. If you doubt this, consider you own habits. Are you professionally inclined to broaden and expand? Or to narrow, limit, and refine?

When Einstein published on the theory of relativity, it was said that fewer than a dozen scientists were qualified to understand what he was talking about. Today, not being understood has become a way of life. Organizations from universities to industries can boast staff specialists who are masters of obscurity. In our looking-glass world, they no doubt earn lucrative promotions.

However, the evidence is strong that obscurity and narrowness can invite calamity. Technically and politically, decisions are being made that affect not only future well-being but even the future survival of the human race. Not being understood and not understanding lead to irresponsible and unsafe practices. Specialized or not, technologists have an obligation to learn how the other half lives. And vice versa. Ignorance has become a short fuse into an atomic nightmare.

Today bridges are vital between those who deal in facts and numbers and those who deal in ideas and words. Workers in technical professions must develop insight into fundamental human values, cultural concepts, and social mores as they relate to technology and to themselves as humans. Application of these values is indispensable to the management of professional careers in the modern world. At the same time, humanists, philosophers, and nonscientists must learn the rudiments of technology. Without broader communication and understanding, trouble will dog the steps of everyone, technologists as well as nontechnologists. Both scientific knowledge and humanistic knowledge have proved too critical for hoarding. The lack of either means jeopardy for all. "As the world becomes more technically unified, life in an ivory tower becomes increasingly impossible," Bertrand Russell wrote in 1960. Russell warned that man could "find himself no longer in the ivory tower, with a wide outlook over a sunny landscape, but in the dark and subterranean dungeon upon which the ivory tower was erected" (3).

The message is plain. To avoid dungeons, we must steer clear of ivory towers in the future, and their scientific counterparts—doorless, windowless laboratories. The traditional residents of each must stop looking exclusively in mirrors.

"Every wall is a door," declared Emerson with sprightly paradox, but only point of view determines whether the paradox is correct. Think of a door as an impenetrable wall, never try opening it, and you make it an impenetrable wall. The reverse is also true. Whenever mankind has achieved a breakthrough, it came from finding doors in previously solid walls. Technologists are prominent among those turning walls into doors, but by no means exclusively. The trick has been turned by artists as well as physicists. Awareness of breakthroughs that liberate the human mind is needed whether the source is a Beethoven, a Goethe, or a Rutherford. "Let us not look for the door, and the way out, anywhere but in the wall against which we are living," advised Albert Camus (4).

Finding doors is one of the first steps. Opening them is next. Then more is essential. Specialists and nonspecialists have to leave their towers and their laboratories. They voluntarily need to cross the chasms that separate them by constructing bridges of cooperation and trust.

Technologists have not taken the lead in the challenge of understanding, but they could. They are professionally skilled in solving such problems as crossing chasms. "Everybody knows that the modern world depends upon scientists," claimed Bertrand Russell, "and, if they are insistent, they must be listened to. We have it in our power to make a good world; and, therefore, with whatever labor and risk, we must make it" (3).

Technology also puts it in our power to destroy a good world, hence the opportunity and responsibility. Astronomer Harlow Shapley considered the challenge optimistically: "Technology and pure science, if opportunity is open to them, can make cultural survival irresistibly desirable . . . Science has had experience in friendliness. It could save us if we gave it full opportunity" (5).

Technological "friendliness" expresses itself repeatedly in international scientific cooperation and tolerance. An attitude of openness could be the greatest contribution of technology to human values and knowledge.

Yet technology in general has not concerned itself with values. Gestalt psychologist Wolfgang Köhler even stated dogmatically that "science does not deal with values" (6). J.W.N. Sullivan also stressed the impartiality of science when he noted that Galileo could seek the law governing the motion of falling bodies without caring how the law turned out. "He could search for the truth with a single mind," wrote Sullivan, "because none of his emotions could be outraged by the result" (7).

The time has come, if not to care how a law turns out, at least to care how it affects the lives of others and the impact it has on earth's biosphere. Facts alone are no longer sufficient to monopolize scientific attention. Human values are now seen as equally necessary in determining truth. If questions of value bring on scientific fear and trembling, that is all the more reason to insist on asking them. Failing to evaluate facts in terms of meaning and value is increasingly recognized as a scientific cop-out.

There are those such as Professor Shapley who believe that science can "save us," yet others are less certain. Wolfgang Köhler for instance:

> In recent years serious doubts have been raised as to whether, in its present course, science will be able to contribute much to the fundamental issues of mankind. Are we to infer that the philosopher and the scientist should live and work each in a world of his own? Actually, no boundaries separate the problems of one from those of the other. Thus, if there be no contact, something must be wrong either with philosophy or with science, or, perhaps, with both (8).

Sponsoring contact among intellectual communities is one purpose of this book, for the sake of technology and those served by it. This won't be easy. Both veterans and recruits in the two cultures must learn one another's language and the common language of human values.

TWO SIDES OF THE CAMPUS

In 1959 a new idea took the intellectual world by storm. It was heralded as new, yet really wasn't new at all. The timing simply happened to be right, and the idea echoed what many men were thinking. With his book, *The Two Cultures and the Scientific Revolution,* English author-scientist C. P. Snow set off avalanches in intellectual circles that have never quite subsided. By coincidence, exactly 100 years earlier in 1859, Charles Darwin published *On the Origin of Species by Means of Natural Selection,* a book that also shook the intellectual world to its eyeteeth and started earthquakes whose rumblings still continue.

C. P. Snow's thesis in 1959 achieved comparable impact. Snow reported on the schism he had noted between two large sections of modern intellectual life, what might be called the two sides of the campus. On one side were technologists; on the other were literary intellectuals, liberal arts people, and their associates. It disturbed Snow that there was almost a total absence of mingling or crossover between the two groups. He considered the split potentially disastrous for western civilization. Each side of the schism was deprived of access to vital knowledge, and the consequences for mankind in general were both intellectually crippling and highly dangerous.

Literary intellectuals, he believed were severely hampered by a lack of understanding about technology:

> But I believe the pole of total incomprehension of science radiates its influence on all the rest. That total incomprehension gives, much more pervasively than we realise, living in it, an unscientific flavour to the whole "traditional" culture, and that unscientific flavour is often, much more than we admit, on the point of turning anti-scientific. . . If the scientists have the future in their bones, then the traditional culture responds by wishing the future did not exist (9).

The situation was discouraging. Incomprehension did not make a healthy outlook for one group, often in control of education, politics, and business, to have toward an equally important group. Snow emphasized the need to teach the principles of science and technology to all students just as history and languages are taught. A student understanding *Hamlet* but knowing nothing of the Second Law of Thermodynamics could hardly be considered adequately prepared for adult decision-making in the modern world.

It was equally valid that the science student with no comprehension of nontechnological subjects had been poorly prepared for modern challenges. When Snow examined the comprehension levels in Britain, with one of the earth's most literate populations, he found little cause for rejoicing. As a scientist recruiting scientists during and after World War II, Snow estimated that he and his colleagues had interviewed approximately 30,000 to 40,000—25 percent of the scientists, professional engineers, and applied scientists in Britain.

> We were able to find out a certain amount of what they read and thought about. I confess that even I, who am fond of them and respect them, was a bit shaken. We hadn't quite expected that the links with the traditional culture should be so tenuous, nothing more than a formal touch of the cap (10).

Snow might have been less shaken if he had paid attention earlier to the point of view expressed by George Orwell in an outspoken 1945 essay, "What is Science?" Orwell examined both cultures, technological and nontechnological, with calm detachment. He was less convinced than Snow that superficial educational programs designed to teach science to the masses would be very useful, especially if they were implemented at the cost of traditional courses in literature and history. The usefulness of scientific education for nonscientists Orwell considered to lie chiefly in "the implanting of a rational, sceptical, experimental habit of mind" (11). If students could acquire a "method" for assistance in finding a solution to any problem that came along, such education would be significant and useful.

But Orwell skeptically wondered if even scientists had much of a "method." He thought from his observations that scientific education in general meant accumulation of facts rather than the development of an experimental approach to problems with a questioning cast of mind.

Snow's contention that technologists should be informed about traditional culture received emphatic support from Orwell. Orwell was dubious of the popular assumption in the 1940s that technological training was excellent preparation for objective judgments on political, social, and moral questions.

> There is not much reason for thinking so. Take one simple test—the ability to withstand nationalism. It is often loosely said that 'Science is international', but in practice the scientific workers of all countries line up behind their own governments with fewer scruples than are felt by the writers and the artists. The German scientific community, as a whole, made no resistance to Hitler. . . The fact is that a mere training in one or more of the exact sciences, even combined with very high gifts, is no guarantee of a humane or sceptical outlook (12).

Orwell considered it imperative for a humane outlook to be developed among technologists. He strongly counseled broader acquaintance with the general culture as exemplified by history, literature, and art. Orwell previewed C.P. Snow's thesis with this sharp 1945 comment on the two cultures.

> A hundred years ago, Charles Kingsley described science as 'making nasty smells in a laboratory'. A year or two ago a young industrial chemist informed me, smugly, that he 'could not see what was the use of poetry'. So the pendulum swings to and fro, but it does not seem to me that one attitude is any better than the other. At the moment, science is on the upgrade, and so we hear, quite rightly, the claim that the masses should be scientifically educated: we do not hear, as we ought, the counter-claim that the scientists themselves would benefit by a little education (13).

Snow was fond of technologists and shaken by how little they knew outside their own subjects. Orwell was the gadfly of his time, fond chiefly of truth, and not in the least surprised, though he also regretted that technology students did not learn enough to be educated citizens by his standards.

Snow pointed out that the split in the intellectual community both reflects and magnifies social problems. It dramatizes the critical question of values. The split automatically establishes battle lines between the "doers" (technologists, businessmen, lawyers, doctors) and the "contemplators" (liberal art students, writers, philosophers, psychologists, sociologists, artists, historians, political scientists).

On university campuses and throughout our society, the result is a tendency for doers and contemplators to be as remote from each other intellectually as the East and West in Kipling' poem. There are impressive exceptions, of course. Some technologists have read widely. Some literati have mastered scientific thought. In general though, the doers and contemplators have little knowledge of or interest in one another. Even a common language for purposes of communication is lacking. Doers continue blithely doing with minimum regard for consequences. Contemplators criticize exhaustively, but offer few practical remedies for the problems they criticize.

Are the doers and the contemplators a twain that will never meet? The answer could be "no" to this question, but only if new directions are sought and taken. These new directions involve a fundamental reexamination of values taken for granted on each side of the schism, as well as those identified exclusively with one side or another. Since World War II, western civilization has undergone a continuing crisis of values. Civil wars, riots, civil disobedience, ideological confrontations have been symptoms of that crisis; and there is no encouraging evidence that they are over. Ironically, the fact of their occurrence is hopeful. Active pursuit of common values is essential as the only means of reconciling men to a common purpose and replacing division with union.

Identifying common values between doers and contemplators is the heart of the problem intellectually, and thus the place to begin. What objectives of society can both cultures accept for the future? Which traditional values will pass through the electronic baggage check for transport to the future? How can the obscuring fogs be penetrated to a communion of values?

With mutual accommodation, the two cultures could begin sharing knowledge as well as values. First though, basic objectives need clarifying to avoid wasting effort pursuing murky goals. Until then, both cultures will continue to suffer the ills of division. That neither side of the campus quite knows something important is being missed simply proves the ailment has not been diagnosed and the patient tries to pretend nothing is wrong. It won't work; such pretense always backfires.

Some technologists know they need to answer George Orwell's indictment by reassessing their objectives and redefining values. From that tentative start, they must practice crawling in the other culture, and eventually learn to walk. Pascal said "a man does not show his greatness by being at one extremity, but rather by touching both at once." Teilhard de Chardin, 200 years later, offered a confirming echo: "To see life properly we must never lose sight of the unity of the biosphere that lies beyond the plurality and essential rivalry of individual beings" (14).

Reconciliation of extremes is one of the oldest values to which men have paid attention through the ages. Achieving it may have failed more often than not, because the objective was nothing less than human harmony, a goal that has proved tougher to reach than the moon, the summit of Everest, and the four-minute mile. If human harmony was an unattained summit in earlier, simpler times, it is an even harder climb now. But perhaps we have learned—or are learning—the penalties of failure if we do not achieve harmony. Thanks to technology, it has become quite feasible for man to end the world with a bang, with no time for T. S. Eliot's "whimper." C. P. Snow offered a hesitant partial remedy:

> Changes in education are not going to produce miracles. The division of our culture is making us more obtuse than we need be: we can repair communications to some extent: but, as I have said before, we are not going to turn out men and women who understand as much of our world

> as Piero della Francesca did of his, or Pascal, or Goethe. With good fortune, however, we can educate a large proportion of our better minds so that they are not ignorant of imaginative experience, both in the arts and science, nor ignorant either of the endowments of applied science, of the remediable suffering of most of their fellow humans and of the responsibilities which, once they are seen, cannot be denied (15).

What Snow suggests we can have with luck and effort has the sound of plenty, because it is so much more than we currently have. Yet it seems as remote and difficult to reach now as in 1959, when Snow wrote. The hope is that we have learned more and can try harder. Man's reach has always exceeded his grasp. (It should, according to Robert Browning.) But now, with the human future depending on it, perhaps both reach and grasp can be extended.

COMPETING UTOPIAS

An English professor and an engineering professor (coauthor of this book), friends and colleagues, scheduled a unique seminar on their university campus. Entitled "Values in a Technological Society," the seminar in their more grandiose visions was seen as a means of eliminating the division between technological and traditional cultures. The seminar would inaugurate a movement that would spread across the intellectual map, radiating light, convincing lions and lambs, engineers and classicists, to lie down together and war no more.

What happened was that the congenial professors almost stopped talking. They discovered with astonishment that they were themselves victims of the schisms. When the time came for intellectual choices in the seminar, they found that their viewpoints diverged on almost everything. They even lacked a common communication system, and key words held different meanings. Their cleavage put them in separate camps and inspired fusillades of argument rather than scholarly peace.

Discussing respective Utopias, the English professor commented, off-handedly but in all seriousness, that his Utopia would not require engineers, including those of his colleague's specialty. The engineer was perturbed since his Utopia had not been designed to exclude teachers of English. The engineer sarcastically pointed out that if engineers were excluded there would be no lights for a library, no way to duplicate books, and the surplus time currently given to study would be needed to till the fields with primitive tools that laymen could manufacture without the aid of technology.

The English professor explained that he had meant all persons would be trained in his Utopia to perform any necessary service. The idea that any man could be his own engineer as well as brain surgeon and automobile mechanic seemed to the engineer equally unreasonable.

The English teacher did not live in a cave. His family did not grow food with basic implements. He worked in a large building packed with conveniences. He relied on secretaries and their electric typewriters. He traveled by automobile, and in countless ways benefited from the efforts of specialists he would omit from his Utopia. He was not a primitivist bent on a retreat to the simple life. He was simply ignorant of technology, fearful of specialists, and vaguely wistful about dispensing with them while retaining their specialties.

Such thinking sounds innocent, but it is not uncommon. It haunts our society from the irrational demands of some environmentalists to the antiscientific biases of certain lecturers using sensitive microphones, perfected by technologists, to denounce technology.

Then there are the technical people, whose numbers are legion, ready to dismiss all aesthetic considerations to concentrate exclusively on technical matters. They display the value standards of those who would ignore a feast table in favor of leftovers, or who would prefer a sunset on television to the real thing.

The engineer's experience with the English professor underscores the division between the two sides of the campus. It accents the importance of real values that have earned acceptance by both cultures.

What are human values? How do they relate to technology? The questions obviously are important to both doers and contemplators. The answers may not be readily apparent, but finding them is essential. Most technologists have been paying an exorbitant price for their absorption in technical specialties, with the values in traditional culture excluded through indifference or neglect. English professors and others might not be ready for reconciliation by addressing themselves without prejudice to the values of technology. But someone must start, and it might as well be from the technology side.

INITIAL VALUES

Every word becomes a value judgment when a definition is attached by usage and attitude. A dictionary will supply the generally accepted core meaning of a word, but if the word is at all controversial, such as technology, the real meaning will vary widely among users.

Obviously, an engineer may *feel* about the word technology quite differently from an English teacher. If they began with the same core meaning, their attitudes would cause a quick divergence. The scope of our value crisis is emphasized by the fact that in the same society, technology will be a scare word to some, a praise word to others. Ironically, those who view technology with apprehension are its direct beneficiaries, with little inclination to abandon the benefits technology has provided in the course of the industrial revolution.

Technology created industrialization and has kept it growing. C. P. Snow realistically admitted that "industrialisation is the only hope of the poor," and added, "The industrial revolution looked very different according to whether one saw it from above or below. It looks very different today according to whether one sees it from Chelsea or from a village in Asia" (16).

Those who don't have it, want it and will pay any price, including revolution, to get it. Some who have it, want the benefits without crediting technology. Understanding these opposed attitudes is necessary to understand the values, both positive and negative, that are applied to technology.

Engineering is a major branch of technology. Ralph J. Smith has defined engineering as follows: "The professional art of applying science to the optimal conversion of natural resources to the benefit of man" (17).

Science is another branch of technology, and science might be defined as systematized knowledge determined by observation, study, and experimentation, as opposed to intuition, belief, or expectation. Although empirical engineering preceded science, the amalgamation of the two has distinguished modern technology. Science and engineering commonly overlap, and differentiation between the two in many situations may be difficult and unnecessary. Scientists and engineers tend to make similar judgments concerning values. Thus in this study, both can be grouped together as technologists.

Technology for our purposes, means the combination of science and engineering, and the multitude of activities resulting from this union. The word technology has curious origins. The combining form "techno" comes from the Greek word for art or skill. Another Greek ancestor of technology is "technologia," meaning systematic treatment. Early relatives of technology, such as technography (description of the arts) and technonomy (practical application of the principles of the arts) were committed by definition and usage to the arts. Our word, technology, and its modern meanings (industrial science, scientific study of the practical or mechanical arts) appeared early and was in general use by the early seventeenth century, as in Tobias Venner's 1628 work on baths aimed at establishing a "baths technologie."

In modern usage, definitions of technology usually include or imply human benefits. Technology "systematizes and applies practical knowledge" for the benefit of man. Because of this traditional tagalong that accompanies technology, the work of the technologist always has a moral commitment by definition, or assumption, if not always practice.

The technologist's efforts *should* improve the human situation. The efforts of many technologists in the recent past were required to develop and manufacture weapons of war. In connection with the atomic bomb during World War II, for instance, debate still rages concerning the morality of technologists' participation. Orwell wrote of British and American physicists who refused to continue doing research on atomic bombs after the devastation at Hiroshima and Nagasaki: "Here you have a group of sane men in the middle of the world of lunatics . . . I think it would be a safe guess that all of them were people with some kind of general cultural background, some acquaintance with history or literature or the arts—in short, people whose interests were not, in the current sense of the word, purely scientific" (18). Orwell had little faith in the ability of scientists and engineers to live up to the obligation of human benefits in extreme situations such as war. Some might argue, and have, that technologists benefit mankind by perfecting weapons that bring an earlier end to hostilities, thus saving lives, as in the case of the atomic bomb. The point is that whether technologists are being lauded or criticized, a moral aspect always exists. They are praised when the moral responsibility of serving mankind is met, condemned when it is not. It is a compliment to the power of technology for both good and evil that the moral aspect is stressed so consistently. This is not accidental.

The reason for moral concern is conspicuous. Technology can both demolish and uplift not just a few individuals, but all mankind, hence the constant question of moral and human values.

Turning to the expression "human values," we first confront the apparent truth that man is the only earth creature capable of making value judgments and carrying out long-term efforts to reach value goals. But then we run headlong into major difficulties, with the word "value."

When questioned, most individuals would insist they have personal values. Some would even claim that their city, state, or nation functions in compliance with certain values, not because of "have to," but because of "want to." Yet in the case of both personal and public values, serious problems arise. Presumably values provide guides to action in life, yet most of us seldom examine our value system very carefully or establish our value priorities logically. We follow norms laid down by parents or peers, and when differing courses of action present themselves, behave expediently rather than in accordance with a design that has its origin in explicitly considered values.

This is typical in a society that values action above contemplation. Expedient action may be the fundamental recipe for survival, but beyond expediency, few individuals consciously map out or revise their value systems, while only cataclysmic events (war, governmental scandal) seem capable of influencing a nation to do so.

Although values constitute guides to action, the action is often not taken despite the prescription given. Kurt Baier, recognizing this fact, stresses attitude rather than action in his discussion of value. A value indicates a favorable attitude, or a behavioral disposition toward the realization of a situation or state of affairs (19). Action may or may not be taken depending on the intensity of the disposition, the scope of the problem, and individual will. With this approach, values become measurable by comparing predisposition with actions taken. The approach is also entirely compatible with the view that we enjoy freedom of choice. Thus, no matter how programmed or conditioned, man still has some "free will." To act or not to act, as with Prince Hamlet seeking justice or Presidents making life and death decisions, is a constant free choice for every man. Whatever action is taken or not taken results from the value judgments and attitudes that precede it. The concept of free will is indispensable to the concept of value, since freedom of choice is essential if value dispositions are to be meaningful in the context of acting or not acting.

In his consideration of value, Baier also includes the factor of amelioration. He attaches the moral injunction that values must lead either actually or theoretically to an improvement in the human condition. The value dictates that man should better his situation, not worsen it. This often backfires, of course. Leaders have strewn history with acts of "improvement" that in the long run turned out to be devastatingly bad. Individuals in their well-meaning but fumbling efforts to make something better have often inexorably fouled up. But the *theory* of improvemnt is nevertheless an inherent aspect of value. It would be a crippling contradiction to think of any value leading to a poorer rather than a better life.

The complete definition of a value, following Baier, becomes:

1. **The value indicates a possible situation toward whose realization there is a favorable attitude.**
2. **When achieved, this realization will bring an improvement in life.**

What is an improvement? What is good? What is bad? The definition of value, not surprisingly, leads to other terms. Schools of philosophy have offered employment to philosophers and remained in session for centuries struggling with the meaning of such terms. For our purposes, we do not have to register in these schools before continuing. The majority of mankind, recognizing that "good" is an abstraction with many variants, has generally agreed certain things are "good." Thomas Jefferson mentioned some in his phrase "life, liberty, and the pursuit of happness." Other "goods" include human security, food, clothes, shelter. If even one of these could be guaranteed to the whole of mankind, the status of man today would be dramatically improved.

As technology increases and widens material prosperity, perhaps such basics will become more generally available. When that occurs, it should be expected that values will shift from these basics to other concerns. Studies of advanced technological societies indicate the gradual occurrence of such value shifts, some of them surprising. For instance, it has been found that computer technology instead of converting men into ciphers and automatons as often predicted may actually have the opposite result. Instinctively wary of herd absorption, human beings may voluntarily choose individualism, freedom of thought, and self-reliance. There are indications of such a trend in areas where technology has aged enough to be taken for granted.

Whatever new circumstances may promote, men will continue to identify with certain "goods," and select goals on the basis of their values. When changing technology, resource availability, and related factors bring about or force a modification of values, goals will automatically be altered. Historical evidence of this occurrence is plentiful.

Technologists have long played an important role and will continue to do so in defining and achieving human values. They would fulfill this job more effectively if they took the trouble to recognize and act on their responsibilities directly, instead of reacting to the proddings of necessity.

The truth should be evident and clearly understood by the two cultures on both sides of the campus. Technology is here to stay. It must be devoted to the betterment of man's lot. If this is not done, man's future on earth is bleak at best. This is the challenge of modern technology—to sponsor values that sustain human life.

REFERENCES

1. Einstein, Albert, "In Honour of Arnold Berliner's Seventieth Birthday," *The World As I See It*, Philosophical Library, New York, 1949, p. 15.

2. Hoffer, Eric, *Reflections on the Human Condition*, Harper & Row, New York, 1973, p. 84.

3. Russell, Bertrand, "The Social Responsibilities of Scientists," from *Science*, February 12, 1960, reprinted in *Modern Essays*, Ed. Russell Nye, Scott, Foresman and Company, Glenview, Illinois, 1963, pp. 249-252.

4. Camus, Albert, *Resistance, Rebellion, and Death*, Alfred A. Knopf, New York, 1961, p. 272.

5. Shapley, Harlow, *Beyond the Observatory*, Charles Scribner's Sons, New York, 1967, p. 198.

6. Köhler, Wolfgang, *The Place of Value in a World of Facts*, The New American Library, New York, 1966, p. 35.

7. Sullivan, J. W., *The Limitations of Science*, The Viking Press, New York, 1933, p. 279.

8. Köhler, Wolfgang, p. x.

9. Snow, C. P., *The Two Cultures: And A Second Look*, The New American Library, New York, 1964, p. 17.

10. *Ibid.*, p. 18.

11. Orwell, George, *The Collected Essays, Journalism and Letters*, Vol. IV, Penguin Books, Harmondsworth, England, 1970, p. 29.

12. *Ibid.*, pp. 28-29.

13. *Ibid.*, p. 30.

14. de Chardin, Teilhard, *The Phenomenon of Man*, Harper & Row, New York, 1965, p. 112.

15. Snow, C. P., p. 91.

16. *Ibid.*, pp. 30-31.

17. Smith, Ralph J., *Engineering as a Profession*, McGraw-Hill Book Company, New York, 1962.

18. Orwell, George, pp. 30-31.

19. Baier, Kurt and Rescher, Nicholas, *Values and the Future*, "What is Value? An Analysis of the Concept," Collier-Macmillan Ltd., The Free Press, New York, 1969.

QUESTIONS FOR REFLECTION

1. In addition to responsibility for the performance and safety of a project, should a technologist be responsible beyond this for indicating the possible effects of a project? Should he confine his concerns to the technical and leave social impact considerations to others?

2. Do the technologists of the United States have a moral obligation to assist nations with limited technology, such as those of Africa and Asia? If given, what form should assistance take? Does the individual technologist ideally wait for his government to take the lead in this matter, and if this is not done, to feel justified in doing nothing?

3. Should technologists take more "humanities and social science" courses? Should such courses be presented differently to technologists?

4. Should nontechnologists take courses in the history and prospects of technology? How should such courses be designed for greatest effect? Would a better understanding of technology benefit English teachers, politicians, voters?

5. How is the individual's value system determined? Are schools responsible for values?

6. Competition vs. cooperation. Which should we stress, if either?

7. What effect has the automobile had since 1930 on values, morality, and the social system? What effect has television had since 1950?

8. What should be done to meet the complaint by employers that engineering graduates in general have difficulty writing a good report or communicating verbally? Can an individual acquire more education when his formal courses are completed, and if so, how?

9. To what extent and how should a technologist become active in local, state, and national affairs?

10. Do technologists compose an international group with some loyalties parallel to or above those of purely national interest? If so, what are these loyalties? Should there by more or less international exchange of technological ideas and expertise?

2. Shapers of the Present

"It's easy to build a philosophy. It doesn't have to run."

Charles Kettering

WHO LEADS AND WHO SHOULD

In the United States and Europe, our society became deeply involved with technology. Few will deny this. But whether or not technology dominates society is more debatable. In political and economic decisions, technologists seem conspicuously absent. Lawyers dominate Congress, for example, and few, if any, technically trained men run for legislative posts.

Corporate Boards of Directors contain only an occasional reformed engineer or scientist. At local levels, technologists seldom are heard at PTA meetings, zoning hearings, or City Commission confrontations. The technologists themselves are responsible for this situation. Most have shown interest only for their own narrow professional niches.

C. P. Snow in his *Science and Government* commented ruefully on this situation. He was concerned that men usually unacquainted with technology secretly make technological choices with immense long-range consequences for mankind. He wanted scientists to have a voice in great affairs and in the making of the large choices. One reason was the need to bring decision-making into the open, and to widen the scope of leadership. Another reason was what Snow saw as the superior historical perspective of scientists.

> For science by its very nature, exists in history. Any scientist realises that his subject is moving in time—that he knows incomparably more today than better, cleverer, and deeper men did twenty years ago. He knows that his pupils, in twenty years, will know incomparably more than he does. Scientists have it within them to know what a future-directed society feels like, for science itself, in its human aspect, is just that. That is my deepest reason for wanting scientists in government (20).

Scientists he saw bringing the gift of foresight to deliberations, and it was foresight that he thought chiefly lacking in the managerial and administrative classes who have dominated leadership positions in government and business.

Technologists have rarely held the levers of power. Such control has traditionally been in the hands of politicians, bankers, attorneys, and businessmen. Often these men have less interest in technology than they do in power, profit, or prestige. Benefits may accrue to mankind, yet frequently too there are serious side effects. Thus, the *uses* of technology require close attention.

Obviously, technologists cannot be blamed for the negative features of the modern world since they have not sat in positions of power. Nevertheless, technologists can be severely criticized for not assuming leadership in their areas of special competence. If they have not taken the trouble to speak up effectively when unfortunate choices were being made, they can be required to share the blame for those choices, through omission if not commission. It is an old, much-abused, but nevertheless true maxim that men who know better have an obligation to see that better is done. "It is not, what a lawyer tells me I *may* do," said Edmund Burke in a March 22nd, 1775 speech to the British Parliament, "but what humanity, reason, and justice, tell me I ought to do."

Humanity, reason, and justice are constantly announcing their needs, and if technologists do not listen, they are unlikely to hear. Another question arises however: Would the technologist's active involvement in public issues improve our condition? The technologist would be likely to prescribe technical solutions. One can rightly inquire whether more of the same will cure technologically fostered ills. A few critics, frightened of technology and what it does to the earth and the earth's inhabitants, would say no. They urge the abandonment of technology and a retreat to the simplicities and self-reliance of earlier times. The fact that without technology the earth was able to tolerate only a tiny fraction of its present human population seems to be one of those facts that critics genially choose not to consider lest they prove upsetting.

The abandonment or even the reduction of technology during the foreseeable future (at least the next millennium) appears highly unlikely. C. P. Snow's arguments, in this light, assume convincing dimensions. Technologists' counsel early in the decision-making process would be helpful. However, those in charge are unlikely to beat a path to their door. Technologists will have to emerge voluntarily. At first they may have to make considerable noise in the marketplace to gain attention.

Captains of industry can be criticized for not involving those with useful technical advice. Yet even more, we must criticize technologists for their lack of interest in large affairs. Whether they speak up or not, the technology that scientists and engineers turn loose in the world often rushes forward like a wild machine. It crushes those who stand bewildered in the path. Critics seize this fact to denounce the machine and the technologists who breathed it into life. The critics have a valid point if technologists do nothing to explain the machine, to advise the bewildered, to assure the machine's proper use. We do not blame the bull in the china shop for the broken crockery, but the one who set the bull free without tutoring him in the niceties.

Technologists must tutor their machines. Equally important, they must instruct potential beneficiaries. Historically they have not always bothered to do so.

THE ROOTS OF TECHNOLOGY

Absorbed in our modern technological problems, we tend to forget that technology had ancient beginnings. Melvin Kransberg has pointed out that man could not have become *homo sapiens* (man the thinker) if he had not at the same time been *homo faber* (man the maker). The first indication we have of man's separation and uniqueness from other animals is through his use of tools. Man's progress from the beginning has been followed and chronicled by his tools, left as surviving artifacts through the ages. It is argued that man might not have walked upright or developed his hands if it had not been for his toolmaking impulse and gradually developed skills. Thus, man created tools, and in the process, tools made and shaped the nature of man.

In his *Life of Johnson,* James Boswell spoke admiringly of Benjamin Franklin, whom some might consider America's first technologist. "I think Dr. Franklin's definition of Man a good one," wrote Boswell, "A toolmaking animal."

The origin of the tool impulse cannot be precisely traced, but we can make a surmise. The earliest men struggled to survive in a hostile environment. They were threatened by the elements, by other carnivorous animals, and by the menace of starvation if they failed to hunt sufficient food. The logical assumption is that man's first tools, therefore, were weapons, to use in hunting, to protect against animals, and also to protect against other men.

The discoveries of Louis Leakey and his wife in the fossil beds of Olduvai Gorge in East Africa have pushed the human chronology back to at least a million and a half years. *Homo habilis,* a tookmaking hunter, left his fossils in the Gorge, as well as fossil remains of animals that *Homo habilis* used as food. *Homo habilis* was small in stature. His brain was no more than half the size of modern man's, but it was large enough to calculate that his hand with a stone in it was more effective than his hand empty.

How and when this discovery was made only conjecture can say. Perhaps in a struggle with another creature, *Homo habilis* in desperation lashed out with a rock and had the intelligence to understand the results. Thus perhaps the first tool was born.

In addition to weapons, the earliest men began the slow process of discovering that not only was a hand with a weapon effective, many hands working in unison with many weapons were immensely more effective. Eventually the astonishing invention of social organization and cooperation appeared, at first mainly for the purpose of hunting large animals. Robert Ardrey in *The Territorial Imperative* wrote:

> Faced by equivalent necessities for successful aggression and successful defense, the hominid band faced ultimate necessities for social amity and co-operation and maximum exercise of primate wit. Are we to wonder that emerging man turned to the biological nation, the defense of a social territory, and a society of outward antagonism to weld his numbers into one (21).

Thus the development of that remarkable, and sometimes devastating tool, the state, began. As for individual men, with tools to aid them in the hunt and later in agriculture, attention could be given to making human life more attractive. First men satisfied the basic drives for food, shelter, security, reproduction, and clothing. Then the groping minds of early men began to recognize additional needs for such abstract values as love, beauty, and knowledge. Man's identification with these values distinguished him from other animals more markedly than before. Values beyond basic ones began altering mankind and differentiating the species more and more from all other animals. Man, for instance, developed a unique sense of humor, a recognition of the ridiculous, and a strange ability to laugh at himself and the universe. "Broadly, it is true that all animals except Man are serious," wrote G. K. Chesterton (22). Laughter was only one of the special sounds that men introduced to the earth. Language and speech also appeared and in time supplied men with unique tools for thought.

While admitting the often harsh record left by early men in their struggle to survive, where we came from does allow a certain pride in where we are. The progress of *Homo habilis* to *Homo sapiens* and the technological innovations that have supplied human needs on a vaster scale than even the generations of the nineteenth century would have thought possible has been impressive by any standards. If earliest man killed his own kind and if he still on occasion does the same, he also now visits Mars to photograph the landscape and he converts deserts into gardens. Technologically he is a success story.

History, as customarily taught, considers battles, treaties, immigrations, and dominant personalities. In broader history courses, economic movements, theories, and forces may even be considered with more than superficial mention. But it is a rare course indeed that undertakes to show how technology has shaped the course of western civilization. A dutiful nod will be given toward the "Industrial Revolution," with perfunctory coverage for the fact that the most lasting and successful of man's revolutions has been waged and is still being won by technology.

Omitting technological developments from history's record does more than simply neglect a crucial part of mankind's biography; it gives a misleading and distorted picture. Failing to take into account the pivotal role played by technology in many of the great forward leaps of mankind results in overemphasis on politics, economics, and individuals, more often shaped by than actual shapers of human destiny. For instance, to explain the American Civil War in terms of political and racial factors without fitting Eli Whitney's cotton gin into the historical equation is nearer fantasy than truth.

Some historians in recent years have attempted to correct such oversights. Historians specializing in technological analysis appear and receive attention. I. Bernard Cohen at Harvard researched and explained in depth the impact of technology on Colonial America. Historian Herbert J. Muller gave credit to technology in *The Uses of the Past.* "The rise of science was the quietest as well as the profoundest revolution in history," wrote Muller (23). Instances of such "quiet revolution" have been apparent, though unheralded, throughout history. They are now being more responsibly reported. One historian has demonstrated how the invention of the stirrup completely altered warfare and thereby the political situation of the time.

Even cursory reflection on man's past shows the intimate relationship between invention and change. The ancient civilization of Babylon and Egypt came into being on a base of agriculture and irrigation. The development of measuring systems became necessary to redefine boundaries after floods. These systems could then be used to introduce the science of astronomy by adding the technical innovation of measurement to age-old observations of the stars.

The Phoenicians combined a knowledge of boat construction, winds, and stars, to establish a trading kingdom in the Mediterranean. To administer that kingdom with any sort of accuracy, records were necessary and a more robust method of making notations than those used earlier. The invention of the alphabet followed.

Descriptive names such as the Bronze Age and Iron Age indicate the importance of technology in the lives of our predecessors. Historically, one technical advance led to another. Ingenious men throughout history, some known, many unknown, contributed to the pattern of developing technology from earliest toolmaking to the present. The impression that technology is an exclusively modern development is obviously incorrect.

In modern times technological change accelerated and began setting a dizzying pace through the marriage of empirical technology and science. It may be that this rapidity of development today is what chiefly frightens the critic of modern technology. The current pace gives little time for adjustment on the part of individuals or society.

Technologists should recognize the legitimacy of such concerns about rapidly growing technologies in the midst of static social institutions. The automobile supplies the most familiar example of technology's irresistible influence. In a few decades, the automobile radically changed laws, family life, urban structure, and much more. Critics may lament these changes, but they are voices crying in a wilderness of cars. The availability, pleasure, and convenience of the automobile dominate and shape life styles. Now, perhaps, new life styles will take shape in connection with the automobile as new factors appear: fuel shortages, higher costs, automobile-caused pollution, parking problems, a lessening of pleasure in bumper to bumper jams.

Technologists may come more and more to take the lead in solving problems resulting from rapid technological changes. Leaving the solutions to time and accident, can be wasteful, costly, and cruel. Again, the automobile is a dramatic example. If technologists rather than businessmen and market analysts had been responsible for the decisions involving automobile safety and fuel economy, for instance, the suspicion is that today's automobile would be considerably safer, fuel frugal, and significantly less guilty of pollution. The trouble is that technologists have concentrated on solving technical problems and have left such matters to others—those theoretically expert in knowing what the public wanted. The public, of course, could easily be manipulated to want what was offered. Typically, the public was offered what would serve most liberally the cause of corporate profits. Technologists, actively involved in this gray area where technology, profits, and human needs come together blindfolded, might have assisted in removing the blindfolds. But there is still the future.

Modern technology has changed man and the world without facing all the consequences. Nevertheless, despite failures, technology has been triumphant. If pollsters surveyed the technologically advanced, they might not find that the majority considered this the "best of times." Yet most would concur that now is better than earlier centuries when disease, poverty, and slavery prevailed. It is not yet apparent as Voltaire wrote satirically in *Candide* that "all is for the best in the best of possible worlds." But it is apparent that some things are better in a world of limitless possibilities, thanks largely to the achievements of technology.

Whatever difficulties technology has introduced (hectic lives, automobile accidents, complexities, tensions), for most people the benefits of technology outweigh the difficulties. This is true today. It was even true when this century began. In the "Mainstream of America" series, Stewart Holbrook described the attitude in 1900, and the contemporary echoes are obvious.

> As the bells and whistles and cannon announced the first moments of January 1, 1900, newspapers going to press carried statements attributed to eminent Americans. Elihu Root was certain that the greatest achievement of the past century had been discovery of the Bessemer process for making steel. But Chauncey Depew reached far back to Benjamin Franklin, with his kite and keys in the thunderstorm who, he said, led to Mr. Edison and his Mazda lamp. Many a lesser figure pointed to the marvelous expansion of the United States from the Appalachian range to the Pacific Ocean. And now the vast void which maps of 1800 had been content to label Great American Desert, or simply Unknown, was threaded with telegraph and telephone wires, with railroads, and dotted with towns and cities (24).

January 1, 2000 is no longer unthinkably distant. When that day of new century resolutions arrives, technology will once more be asked to turn dreams into reality. Jacob Bronowski in his 1974 public television series, "The Ascent of Man," repeatedly made the point that technology historically created the conditions that render large populations possible, and that it sponsored tolerable conditions of existence for those populations. Before the application of technologically manipulated power to the problem of survival, most men were serfs, and the idea of democratic government was entertained only as a Utopian concept. Technology in several specific ways freed the serfs in those areas endowed with technology. In the U.S. of the 1970s, men may commit suicide from frustration and disappointment with the "good life," but technology sees to it that few if any of them starve to death. And despite disappointments and frustrations, few living now would willingly consent to return to the difficulties of pre-technological ages. When men speak and dream of the good old days, they don't mean the old days of *Homo habilis,* when survival meant holding a sharpened stone rather than a steering wheel, a calculator, or a book of poetry.

If the pollsters objectively fed both the good and bad of contemporary technology into their computers, they could very well be informed by the emerging tape that *these* are the good old days. In relation to our ancestors, we have come a long way, and virtually none of us want to go back.

REFERENCES

20. Snow, C. P., *Science and Government,* The New American Library, New York, 1962, p. 73.

21. Ardrey, Robert, *The Territorial Imperative,* Delta Book, Dell Publishing Co., New York, 1966, p. 263.

22. Chesterton, G. K., *The Uses of Diversity,* Dodd, Mead and Company, New York, 1921, p. 1.

23. Muller, Herbert J., *The Uses of the Past,* Mentor Book, New American Library, New York, 1954.

24. Holbrook, Stewart H., *Dreamers of the American Dream,* Doubleday and Company, Inc., New York, 1957, p. 255.

3. The Odd Values of Man vs. Man

"The modern community stands at a half-way house. It knows too much to be content with the old ways; it knows too little to establish ways of its own. It attempts to combine the science of the new era with the small rivalries and petty ambitions of the old, without the limitations to which these were subject."

Stephen Ward
The Ways of Life

EFFICIENT TOOLS FOR DESTRUCTION

Of all the maladies that plague man, war has easily been the most horrendous and catastrophic. Earthquakes, volcanic eruptions, hurricanes and tidal waves have destroyed large numbers of people in sudden upheavals. Epidemics have run devastating courses through communities, nations, and continents. Yet since these natural disasters are made in nature rather than in the minds of men, their horrors seem incomparable with the horrors of war. The reason may be that natural disasters impress us as inevitable, while the disasters of war do not. "Made in Heaven" is a different moral label than "Made by Man."

Technology has not made wars. Human fears and ambitions and unbridled nationalisms have done that. But technology has always provided the essential tools of war. In the twentieth century, technology simply provided more efficient tools than before; yet historically the pattern, since ancient days, has been one of constantly increasing efficiency in man's ability to kill. Can technology be blamed for the fact that there have been men in every age quite willing, even eager, to cultivate this ability?

Fear of science in this respect is by no means new. Nearly a century and a half ago, in 1830, Charles Lamb, the English essayist, wrote to his friend George Dyer: "Alas! can we ring the bells backward? Can we unlearn the arts that pretend to civilize, and then burn the world? There is a march of science; but who shall beat the drums for its retreat?"

There have always been those such as Lamb (inspired name for a pacifist) willing to beat the drums for a retreat from war, but relatively few have marched to that wistful tune. Traditionally, the war drums have effortlessly drowned out the peace drums, and the soldiers have picked up the new weapons provided by technology as they marched off to kill this day's enemy, who was yesterday's friend and if he survives will no doubt be tomorrow's ally. "We are a peace-loving people," the national leaders announce on all sides, "So watch your step, buddy. We will kill to keep the peace!"

Technology it sometimes seems has always been more effective in providing the tools of war than the tools of peace. This may be true because it is obviously easier to blow up 100,000 people than to feed, clothe, employ, and inspire them. The French leader Georges Clemenceau admitted after World War I that "the art of arranging how men are to live is more complex than that of massacring them."

In World War II, modern technology outdid itself in the art of proficient massacre. The forces of the combatant nations were given extraordinarily potent devices, and they used them with efficient zest.

The nuclear age began in the summer of 1945 with two convincing demonstrations of technological prowess. An atomic bomb dropped on Hiroshima killed 100,000 people and condemned many more thousands to mangled bodies and lifelong deformities. Another bomb dropped on Nagasaki underscored the fact that technology had introduced a devastating new reality into the war business.

These two atomic bombs were dropped to help end the war, and they succeeded in doing so. After the war, however, progress in destructiveness continued. The atomic bomb has been multiplied in effectiveness by a factor of 500. Many bombs are now available virtually at an instant's notice on intercontinental ballistics missiles and in long-range aircraft that can achieve explosive effects of 10 megatons (equivalent to ten million tons of TNT). Even greater explosive power is available, if any point existed in achieving it. With the numbers of such bombs running into the thousands, the threat to human safety has for many become a question of human survival, as the technology for manufacturing these bombs proliferates among the nations of the earth. If these bombs were used, conservative estimates agree that casualties from the immediate explosion and subsequently from radiation would be so great, the survival of civilization in any country subjected to nuclear war would be doubtful at best.

In 1974, it was reported that the U.S.S.R. has 11 bombs and the U.S. has 36 bombs for each major city in the opposite country. According to the report, each country has approximately 200 cities with populations exceeding 100,000 (25).

While the size and efficiency (in terms of mass-kill potential) of the bombs increased, delivery systems also were advanced by technology. The delivery time shrank from 20 hours for the 300 mile-per-hour B-29 to the 30-minute flight of the ballistic missile. The accuracy of control and guidance systems has correspondingly advanced. Both the U.S.S.R. and the U.S. admittedly have possessed the bombs and delivery technology for a number of years sufficient to destroy each other many times over. In 1969, it was calculated that nuclear explosives equal to six tons of TNT were available for every human being. In the intervening years nuclear armaments have increased faster than population, thus the explosive equivalent per person has also gone up. During the 1970s, the U.S.S.R. and the U.S. seemed rather more serious than before in their recurrent discussions on disarmament and moderating the atomic bomb race. But as noted, at the same time other nations were either in the race or technologically capable of entering it. The spectre of the future was that no nation with any sort of technological base would lack its own defensive/offensive store of atomic bombs.

At the opening of the atomic era, U.S. military policy began naively on a theme of dominance and the illusion that a period of Pax Americana was possible while the U.S. had a monopoly of atomic bombs. There is no way, of course, for a nation to monopolize technology. The Russians soon developed the bomb independently, and other countries have since done the same.

For many years, however, only the Americans and the Russians possessed both the bombs and the technology of massive delivery. And while this was true, a sort of "Pax Terror" existed. The peace of terror continued since one side knew that the other side could and would destroy its destroyer. While this situation prevailed, it was evident that technology had finally provided a tool of war so devastating a nervous but continuing peace was maintained.

What will happen with many nations, more and more of them, sharing the Pax Terror is anybody's anxious guess. The thesis is inescapable that some of those nations will have less to lose than the U.S. and the U.S.S.R., and that national hatreds may be greater than those achieved by the Americans and the Russians. Will such nations resist the temptation to use their bombs? Will the fact that a despised neighbor can retaliate in kind prove convincing? Will "deterrence" successfully deter as it did between the Russians and Americans? The historical record does not encourage optimism on this question. The rationalization that a quick and destructive atomic move will nullify an enemy's ability to respond in kind is easy to make and perhaps impossible to refute when destructive emotions have an inclination to run amok. When the national passions take over, condemnations of war, as in this indictment by St. Augustine, immediately become unpopular: "What do we condemn in war? Is it the fact that men are killed who all one day must die? Only cowards would bring this accusation against war. What we condemn is the desire to harm, the implacable will, the fury of reprisals, the passion for dominion." A nation aroused seldom if ever can forego this desire, fury, and passion. Further, the citizens of a nation are notoriously prone to take up their machine guns and march across borders when asked to do so.

"We must renounce war not because it is terrible, but because it is evil," wrote H. M. Tomlinson. "There never was a good war or a bad peace," concluded Benjamin Franklin, near the end of the American Revolution. These may be recurrent peace drums, but they have a muffled sound now as they did in 1935 when Tomlinson wrote and in 1783 when Franklin wrote. Weapons with a frightening technological effectiveness have maintained a tense and troubled peace since 1945 (let's not count murderous conventional squabbles such as Vietnam, the Middle East, and Northern Ireland). But how long? H. M. Tomlinson wondered much the same in 1935 and World War II began four years later.

> . . . if we did feel bound to speak, to whom and to what could we appeal? A nation's honor? But the march in step of multitudinous feet has no concern with honor. You might as well appeal to the rising tide, or the spread of influenza. To appeal to such things is absurd. A nation has no honor; racial needs, and impulses know nothing of it. Honor, like reason, would but confuse instincts that are on the way to the gratification of a viewless appetite. A nation knows no morality, but only what it wants. Or let us say that the morality by which a nation is guided is shown in those bloody shreds and tangles hanging on that machinery in London after an air-raid. It is there. It comes down to that (26).

Technology complacently has gone on perfecting and providing the pragmatic hardware. Missile technology encouraged the development of anti-missile technology. The vulnerability of aircraft carriers and land bases to missiles encouraged development of missile-launching submarines. And if submarines could dispatch atomic weapons why shouldn't they also be fuelled by the atom, allowing them to roam about unseen in the silent depths of the oceans until the "order" came? So the Americans had first the Polaris submarine and then the Trident. The Russians followed tit for tat.

In another technological step, the U.S. developed the MIRV (multiple-independently-targeted reentry vehicle). This introduces the economical concept of the multiple warhead, several nuclear bombs launched with a single missile and each bomb individually addressed (though not R.S.V.P.) to a different target upon reentry into the atmosphere. Russia countered with a MIRV. Next the U.S. focused technology on development of the MARV, a missile with a more sophisticated guidance system making possible greater accuracy. It made good economic sense to assure that our bombs killed precisely those we wanted them to kill instead of being wasted on Siberia or Idaho.

"Back to the drawing board," said the Russians or the Americans, depending on who was first. And the dance went on, oddly reminiscent of the Lobster Quadrille in Lewis Carroll's *Alice in Wonderland:*

> "Will you walk a little faster?" said a whiting to a snail,
> "There's a porpoise close behind us, and he's treading on my tail.
> See how eagerly the lobsters and the turtles all advance!
> They are waiting on the shingle—will you come and join the dance?
> Will you, won't you, will you, won't you, will you join the dance?
> Will you, won't you, will you, won't you, won't you join the dance?"

Or is the arms race more accurately described as "All of Us in Blunderland?"

WHERE WE ARE NOW

The U.S. and the U.S.S.R. circle each other warily in the manner of sword dancers. They hold disarmament talks. They use the rhetoric of detente. And both sides continue unabated investments in the technology of war.

At the same time, the French, the Chinese, the Indians equip themselves with atomic bombs . . . "just in case." Other countries, such as Israel and Egypt, are thought to have the bombs or to be capable of building them. In the multinational suicide scramble, security decreases as armaments increase. And no one has come within shouting distance of a practical solution to this proliferation.

A new concern has entered the picture in the 1970s: nuclear enrichment. To use nuclear power for peaceful purposes, such as energy, nuclear fuels (i.e., uranium) must be enriched in a complicated process for employment in light-water reactors. The technology of nuclear enrichment would be difficult for most nations to develop on their own. Atomic bombs are much simpler to put together. However, the U.S. has shown occasional willingness to export nuclear enrichment capability to its friends—selected friends—around the world. This policy has been criticized, and may be reduced or abandoned, for one key reason: nuclear enrichment facilities can be used to produce Plutonium-239, the "trigger" for fission bombs. The proliferation of bombs would remain on a limited scale if the nations making them had to purchase or steal the necessary ingredients. Nuclear enrichment plants would greatly simplify the problem and make bomb proliferation a matter only of desire and will.

Does the average human being understand the atmosphere of risk in which all men live beneath the Damocles sword of the atomic possibility? American citizens, including American scientists and engineers, generally seem oblivious of or disinterested in considering such risks. The usual reasons for indifference prevail: 1) habit and inertia, 2) ignorance of the effects of atomic war and a wish to believe they have been exaggerated by critics crying wolf, 3) boredom, apathy, and accommodation from living with the atomic monster so long, 4) blind hatred of communists and other "enemies" of our way of life, and 5) nationalism.

> The limp apathy that we see at elections, the curious indifference in the presence of public wrongs and horrors, the epidemic of sneaking pilferage, the slackening of sexual self-control—all these are symptomatic like the furred tongue, subnormal heat, and muddy eye (27).

When C. E. Montague wrote the lines just quoted in 1922, he warned that "the future is said to be only the past entered by another door." He put his frail hopes for human sanity in the combatants who had lived through World War I. "War hath no fury like a non-combatant," he noted contemptuously of stay-at-home statesmen who would inflict the "red rain" of war because their own noses had not been rubbed in its hideous reality. Montague's advice was to moderate nationalism and support the League of Nations. His advice, of course, was neither heard nor taken, and the second of this century's World Wars arrived in less than two decades.

What will happen now? History provides little reassurance that atomic war will be avoided. All the symptoms are chronic indicators of infinite trouble farther on.

Countries, the U.S. a leader among them, continue unabated preparations for war. The fact that these preparations may be diminishing national security and lessening mankind's prospects for survival is a subtlety yet to be embraced, or acted on. For instance, the U.S. has made itself number one in military power and technology. But the price tag on this eminence has always been high. Although one of the richest countries in the world, the U.S. ranks eighth in the ratio of doctors to patients. It is 14th in literacy and infant mortality, and 25th in life expectancy among the world's nations. Critics of military spending argue that much money spent on military gear is superfluous in terms of realistic defense against potential aggression. They have not advocated elimination of the military (only a Messiah would do that, and he undoubtedly would still be crucified), simply the use of the wasted money for public needs. One helicopter costs as much as a public health center. U.S. military aid to the dictatorship in Greece in 1971 was sufficient to build four 300-bed hospitals. One nuclear powered aircraft carrier costs as much as 62 high schools (28).

It is estimated that the U.S. has spent more than a trillion dollars since World War II and that most has been spent on equipment now obsolete. The U.S.S.R. and other countries have paralleled this spending. Will you, won't you, will you, won't you, will you join the dance? Like the creatures in *Alice in Wonderland,* no one seems able to stop joining the dance. And war values retain their odd appeal.

CAN TECHNOLOGY STOP THE DANCE?

Must the dance of death continue until the human race does itself in? Must the red rain of human blood fall and keep falling? The gloomy truth is that "probably" seems more convincing than any other current answer.

If the past is prologue, as Shakespeare suggested in *The Tempest,* there are no persuasive solutions other than idealistic ones such as Montague's advice that nationalism should be sacrificed in favor of internationalism. Another idealistic answer would be that technologists should "study war no more." If technologists concentrated entirely on the great human problems and on projects for the well-being of man, war machines would sputter, run down, and rust. Nations would have to embark on other hobbies than militarism if technology, bag and baggage, walked away. Then Ed McCurdy's strange dream might become an even stranger reality.

Last night I had the strangest dream,
I've ever dreamed before,
I dreamed the world had all agreed
To put an end to war.

It may be that some technologists will study war no more. Some may sing McCurdy's ballad. But many technologists will continue to work on projects that serve the military needs of armed and frightened nations. Furthermore, it is careless to forget that there are profits in weapons for stay-at-home warriers.

Devoting the technological energies invested in destructive pursuits on fulfilling man's greater human needs would undoubtedly bring remarkable breakthroughs, just as the Manhattan Project organized science in World War II to produce the atomic bomb. Yet, Jerome Wiesner of M.I.T. and Herbert York of the University of California, San Diego, writing on the test ban treaty in 1956, reviewed the arms race and concluded that, "It is our considered professional judgment that this dilemma has no technical solutions" (29).

The melancholy conclusion of Wiesner and York more than a decade later still seems accurate. In this area, technology offers no sanctuary, no way out, no promise of deliverance from the threat of war. Militarism is a symptom of nationalism. Nationalism in turn has its roots, not in technology, but in human social groups and relations. Though technology can make wars murderously efficient, it does not start nor can it be relied on to prevent them.

This does not mean technologists cannot contribute indispensably to the prevention of war. For one thing, they can tell the truth loudly and clearly concerning the destructiveness of the weapons supplied by technology. They can deny individual citizens the psychological illusion of security by throwing the harsh lights of reality in their faces, with facts and figures that emotion can shout down but that reason pushes immediately to its feet again.

Technologists can counter the hobnail values of war with the humanistic values of life, international understanding, and peace. They can force their way into the continuing debate between those who make a holy cause of nation and those who make a holier cause of man. They can question the motives or the good sense of those who prattle of peace while investing half their goods in the paraphernalia of war. They can point out repeatedly that though we have inched out into space as far as the moon and Mars, earth is still the only home man has and the only home he is ever likely to have.

Some technologists will do this . . . some are doing it. Their drums are heard, questioning, challenging drums that don't buy the old ideas. Russian as well as American technologists have been contributing in recent years to the great debate. Their voices afford a modicum of encouragement. Perhaps it does make sense that there is still time, because there is still tomorrow. But it bears remembering that it is late in the day, today.

Because of the hour's lateness, technologists' contributions to the dialogue of sanity are overdue. Whether they will put their experiments aside for a time and speak their minds in favor of human values rather than war values is another of the unanswered questions for historians of the next millennium to answer pedantically in their footnotes—If there are any historians in the next millennium, or if there is a next millennium for man.

REFERENCES

25. *Newsweek*, September 23, 1974, p. 62.

26. Tomlinson, H. M., *Mars His Idiot*, Harper and Brothers, New York, 1935, pp. 71-72.

27. Montague, C. E., *Disenchantment*, Brentano's, New York, 1922, p. 255.

28. Data from *SANE*, 318 Massachusetts Avenue, Washington, D.C. 20002.

29. Wiesner, J. B. and York, H. F., *Scientific American*, Vol. 211, October, 1965, pp. 27-35.

4. People and Hunger

"Be fruitful, and multiply, and replenish the earth, and subdue it: and have dominion over the fish of the sea, and over the fowl of the air, and over every living thing that moveth upon the earth."

Genesis
Chapter 1, Verse 28

THE MALTHUSIAN THREAT

The number varies, but the conservative estimate is that there are a billion and a half hungry people in the world. No, they are not dieting in the pursuit of sleek shapes. They don't have enough to eat, and have no way to obtain it. Every day several thousand will starve to death. In China during the nineteenth century approximately 100 million people starved to death. The arithmetic of hunger can offer countless startling figures.

The chronicles of hunger also offer vivid human scenes, at once humorous and poignant. Writer Aldous Huxley wrote of an incident in India, which dramatically emphasized a particular human value in that land of large populations and large hunger. The focus of the value: manure.

Huxley was riding an elephant in the Jaipur region. Suddenly in the street, the beast saw fit to relieve itself on the mammoth scale possible with the larger mammals. An elderly peasant woman ran joyfully with a container from her nearby shack. She eagerly took possession of the elephant dung. It was a commodity to treasure. When dried, the dung would fuel her cook fire several days, and she and her family were removed a little farther from the continuing threat of starvation.

In 1798, Thomas Malthus stated the Malthusian doctrine that population increases in geometrical ratio while human subsistence increases in an arithmetical ratio. This mathematical disparity, if population were unchecked, meant in Malthus' words, "the perpetual struggle for room and food." Further, it meant that vast numbers of people would overwhelm available resources. The result: widespread starvation, upheavals, disorder. Malthus had no solution, other than to check the growth of populations, which he did not expect to happen. He predicted disaster.

It did not come right away. The Malthusian doctrine was widely challenged when it was first announced and during the century that followed. The theory received attention because it was bold, dramatic, and controversial, but for a long time it seemed incorrect at least for underpopulated territories such as the Americas. The virgin lands of North America for more than a century efficiently absorbed much of the surplus populations of Europe. Industrialization and a series of wars kept European and American population growth further in check. European surpluses also found havens for immigration in other continents—Africa, Australia, Asia.

With the multitudes of India and China out of sight and out of mind, it is small wonder that Malthus' thorny predictions bore no fruit. In the twentieth century's second half, however, the Malthusian doctrine receives less contempt and fewer jeers. Several things happened. The American frontier vanished, as did those of other continents. The average life span increased not only in the industrial nations, but in other nations as well. It was during the second half of the twentieth century that William Vogt asked an aged man in India what the greatest changes were in his own lifetime. The first was independence, but the second was: "The coming of the doctors." Then the old man added: "Of course when I was a boy we often had three meals a day. Now there are more mouths to fill and we do not always eat twice" (30). The irony was unintentional but inescapable. The doctors preserved life; more life increased hunger. Thus, a deluge of people brought a flood of hunger.

Today the starving multitudes of India, China, Africa, and South America are no longer out of sight and out of mind. They are on the 6:30 television news. They haunt the front pages of daily newspapers. And their hunger sounds can never quite be shut out by turning up the stereo. The arithmetic of hunger coupled with shortages of energy seem to be putting everyone on the tracks of elephants.

For a considerable period it was thought that Malthus erred. There were food surpluses in many areas, and with mechanized agriculture plus suitable fertilizers, it was hoped that food-plenty could be made a worldwide fact. However, world population growth began exceeding agricultural growth, particularly in the hungry countries. Hunger instead of controlling population growth served to stimulate it. Thus, in the twentieth century, the affluent countries of the West came in time to stable or even declining populations, while the hungry continents multiplied at a frightening pace. There were various technical reasons, such as an apparent correlation between insufficient protein and fertility, the tendency of nature to preserve a threatened species with a higher birth rate, and the historical pattern in which large families and poverty have gone together.

Whatever the reasons, the Malthusian doctrine showed signs of vitality in the second half of the twentieth century. The threat of disaster was not apparent for the affluent, that fortunate 20 percent of the world's population controlling 80 percent of the world's wealth (31). But the question of the age was the ancient one between the rulers and the ruled, the wealthy and the poor. Could the 20 percent enjoy its feast with the 80 percent starving outside the stockade? Scientist Isaac Asimov said, "there is no safety for some while everyone else is in misery." And the lessons of history contain many stark warnings in this connection.

> It was the bad harvest of 1788, with resulting scarcity and high prices of food in Paris, that precipitated the frenzy of the French Revolution. A mob of women set off for the Assembly to demand bread. They were joined by men. The members of the Assembly fled, and the mob turned and went to tear down the Bastille, the symbol of the power of government (32).

The disaster of uncontrolled population growth and hunger inevitably seems to affect all passengers on spaceship earth, even the first class passengers, though the trouble begins in steerage and guards with guns are stationed at the ladders leading to the upper decks.

The dimensions of the problem are staggering. A rate of population increase of only 1 percent each year means a doubling of population every 70 years, while a doubling occurs in 23 years if the rate of increase is 3 percent, a rate now exceeded in some parts of the world. The average world increase is now about 2 percent (33, 34). The world's population in 1800 was about 0.7 billion; 2.5 billion in 1940; and in 1976, the 4 billion mark was exceeded. By the year 2000, a world population of nearly 7 billion is anticipated.

There are approximately 70 million more mouths for the human race to feed annually. This requires 30 to 40 million new acres in food production just to keep the human race at its current level of malnourishment and hunger. The grim agricultural and economic facts of life are that vast portions of the earth's surface can never produce sufficient food to support the populations already inhabiting them. It is estimated that 30 percent of the earth's surface, or 17 million square miles, is cultivable. Yet much of this land would need massive technological efforts as well as chemical fertilizers, recently in chronic short supply, to become significantly productive.

The numbers tend to overwhelm credulity. India has five million new people to feed each year. To do so it would have to acquire 700,000 additional tons of cereal annually. India for many years has been nowhere near self-sufficient in food, depending on other countries such as the United States. Or it starves.

It took mankind its first million years to reach a population of three billion. That figure will be more than doubled in less than half a century. What then, if not hunger, does the future hold?

THE TECHNOLOGICAL HOPE

Paul Ehrlich in *The Population Bomb* has said that it is already too late to save millions from starvation (35).

Our standard of values is brought into question by this situation. Disagreement has arisen concerning the seriousness of the problem, and general inertia has limited the amount of an effective attack. Reactions have ranged from the innocence of Marie Antoinette's perhaps apocryphal "Qu'ils mangent de la brioche" (if the people have too little bread—let them eat cake) to the prickly realism of Senator Strom Thurmond, on the hunger problem, "You had them back in the days of Jesus Christ, you have got some now, and you will have some in the future. You will always have some people who are not willing to work;" to Barbara Ward's concern that earth is "the lunatic asylum for the planetary system" when she compared annual expenditures of $150,000,000,000 on military concerns with the spectre of hunger ubiquitous in the world (36).

With a majority of mankind hungry, Barbara Ward asked a question that so far at least has been answered negatively with inaction: "Should we extend our vision to include all the peoples of our planet?" Her answer was this:

> To be content with anything less is to miss the whole scale of the moral challenge of our day—which is nothing less than the ability of our civilization to redeem its wealth, turn its immense resources to the service of life, and use the technology of abundance to recreate, not destroy, the face of the earth (37).

The technology of abundance has certainly an abundance of challenges in this area. How can population growth be controlled? How can erosion be stopped and land restored to productivity on a scale unimaginable before in human history? How can an agricultural revolution be achieved in areas currently tilled with the methods of Biblical times? If the death rate is lowered in an area, how can the birth rate be prevented from creating conditions of untold hardship and human pain? How can the nations with 80 percent of the world's wealth be persuaded that it is in their interest to recognize the common fate of mankind and to help those who have no wealth?

Making land substantially more productive would counter Malthusion predictions of disaster. But can production increases keep ahead of the exponential increase in population? In parts of the world, they have not. In other parts of the world, the time when they will not is visibly approaching.

It is said that more land can be settled and brought into cultivation. However, areas with the worst food shortages and the greatest population pressures often are already over-cultivated, and the land is exhausted by poor farming methods and unceasing use for centuries. To restore the land would require intensive and costly effort. Cultivating new lands is a continuing struggle putting pressure on oil and gas reserves for both energy and the production of essential fertilizers.

The argument is presented that earth's low population density can be increased, but there are areas already critically overcrowded. Increasing the populations further would merely increase strains, social unrest, the prevalence of crime, and the diseases of anarchy or revolution (38).

Some say that the oceans can be cultivated to provide sufficient food for enormously larger populations. This optimistic dream ignores the sad reality that oil spills, garbage dumping, human pollution, and long neglect have made the oceans threatened areas too. Certain whale species have been hunted to the point of near annihilation. The "cod wars" in the waters around Iceland are further proof that the oceans are already being exploited to near capacity in many areas. The fact that industrial pollutants have been found in the waters of the Southern Ocean around Antarctica is not an encouraging indication that the oceans will be ready whenever man turns to them for his next meal. In desperation he may turn to the oceans and discover that he has made them sick as well as himself.

Perhaps the last romantic illusion of beleaguered man is that if worst comes to worst, he can always immigrate to another planet orbiting another star. With this frail pipedream for consolation, he seems to accept the progression from worse to worse with no serious effort to change his habits or mend his ways.

Still, it is true that a technology of abundance has worked wonders in the U.S., Europe, and other parts of the world. Even when the bleakest side of every problem is admitted, technology can still offer enormous help if those with the technology available possess the will to help. It becomes, in short, a question of human values. It would require overcoming what Barbara Ward meant when writing, "the nationalist framework of all our training keeps the poverty of the world at large outside our vision, beyond the reach of our imagination, and far away from any commitment of justice and good will" (39). It would require leaving behind the idea patriotically expressed by a member of the American delegation to the 1974 World Food Conference in Rome to the effect that the world must not expect the U.S. taxpayer to foot the bill. It would require a more active generosity and tolerance than heretofore displayed, or at least a prudent realization that when one turns his back on the privations of others he risks a knife in the back. Economist John Maynard Keynes warned about this shortly after World War I:

> Men will not always die instantly. For starvation, which brings some to lethargy and a helpless despair, drives other temperaments to the nervous instability of hysteria. . . These in their distress may overturn the remnants of organization, and submerge civilization. . . This is the danger against which all our resources and courage and idealism must now cooperate (40).

In other words, if human values are not sufficiently persuasive, it may be in our own best interests to help with whatever technology we have available, even perhaps to break our loaf in half and give part to those who have none.

REFERENCES

30. Vogt, William, *People! Challenge to Survival*, William Sloane Associates, New York, 1960, p. 4.

31. Ward, Barbara, *The Lopsided World*, W. W. Norton & Company, New York, 1968, p. 11.

32. Boyd-Orr, John, *The White Man's Dilemma—Food and the Future*, British Book Centre, Inc., New York, 1955, p. 30.

33. Davis, Kingsley, "Population," *Scientific American*, September, 1963.

34. Hardin, G., *Science, Conflict and Society*, W. H. Freeman & Co., San Francisco, pp. 101-110.

35. Ehrlich, Paul R., *The Population Bomb*, Ballantine Books, Inc., New York, 1968.

36. Ward, Barbara, p. 74.

37. Ward, Barbara, p. 101.

38. Calhoun, John B., "Population Density and Social Pathology," *Scientific American*, February, 1962.

39. Ward, Barbara, p. 88.

40. Keynes, John Maynard, quoted in "The Politics of Hunger: II," Anthony Lewis, *The New York Times*, October 31, 1974, p. 41.

5. Energy, Matter, and Pollution

"We are going at top speed and we are using all our natural resources as fast as we can. . . But when our resources run out, if we can still be ahead of other nations, then will be the time to brag; then we can show whether we are really superior."

Will Rogers

APPROACHING THE END OF OIL

Crisis came to the American people in 1973 and 1974. The service stations displayed signs with ominous words: "No Gas." Middle Eastern suppliers had imposed an embargo. The supertankers were stopped. The American motorist was threatened with a grisly prospect—walking—as long lines formed at stations. Hydrocarbon energy was hoarded, fought over, and anxiously protected. Fears developed that the tank would be emptied and never filled again, so motorists lined up at stations to buy a quart of gasoline and top out an already full tank. The paranoia of man's energy future was tried on for size, and it was a tight fit. "I don't like the looks and feel of it," decided the American motorist, "Bring back the infinite oil."

Some were convinced that the "oil crisis" was engineered by the Seven Sisters (the major oil companies) to justify "highway robbery" in the form of higher gasoline prices.

Finally the embargo was lifted. Tankers resumed their voyages and methodical polluting of the world's oceans. The stations had gasoline to burn again. American drivers put aside their anxieties, turned with relief once more to their gas guzzlers, and began devouring gasoline as if there were no tomorrow that could be counted on.

The truth is, of course, that oil isn't infinite. Sometime tomorrow there will be another oil crisis—an out-of-oil crisis. Mankind's long-range energy prospects are complex, enormously challenging, and innately frightening to people intent on maintaining energy-wasteful habits. Oil is a depletable resource. Oil supplies can last no more than a few decades. Coal is not an infinite resource, though coal reserves are more plentiful than any other fossil energy. Finite uranium reserves indicate that an end of uranium is also visible on time's horizon.

Mankind faces a genuine "energy crisis." Intelligent choices have to be made today, not tomorrow, if energy calamity is to be avoided. Alternate energies, such as solar, wind, tide, geothermal, ocean thermal, and biogas, must be studied and mastered.

But in the United States, the motorists cheerfully fill their tanks and drive away, choosing to pretend there is no tomorrow, and to leave energy concerns to government, the experts, or the technologists.

The United States with about 6 percent of the world's population consumes more than 1/3 of the world's energy and mineral production. The per capita use of energy in the U.S. was the equivalent of 42 barrels of oil per year in 1955. Consumption increased to 60 barrels per year in 1970, with 95 barrels per year forecast for 1985. The total U.S. demand for energy in 1970 was approximately 67 quadrillion BTU. In 1985, the forecast is for 115 quadrillion BTU.

In the 1970s an announced goal of the U.S. by the 1980s was "energy self-sufficiency." When these announcements were made the U.S. was importing almost 10 million barrels of oil daily. A report by the Mobil Corporation, published in *The New York Times,* estimated that by 1985 the equivalent of 35 million barrels of oil would be consumed daily in the U.S., with 26 million barrels imported. These figures have convinced many experts that "energy self-sufficiency" simply provides a political catch-phrase rather than an even remotely achievable goal. Nuclear engineering professor David J. Rose has written that despite the inclusion of all possible new oil sources (four million barrels per day are expected from North Sea and Alaskan wells by 1980), plus synthetic oil from coal, the U.S. would still have to import about 15 million barrels of oil per day in 1980 (41).

The investment to achieve energy self-sufficiency by the year 2000 would be astronomical. Rose estimated that producing six million barrels of oil per day from coal would require an investment in excess of $50 billion.

Known oil reserves in the U.S. stand now at approximately 700 billion barrels. More will be added as the search for the penultimate drop continues. Oil is mankind's energy savings account, on deposit for approximately 40 million years. We can assume that eventually every portion of those ancient solar energies will be withdrawn and used.

In 1976, *The New York Times* reported oil explorations taking place in the Antarctic, with promising results.

> The research ship Glomar Challenger has found evidence suggesting oil and gas beneath the Ross Sea. But since the ship was not equipped to cap blow-outs, it could not drill extensively, so the actual extent of any oil or gas is only conjectural (42).

It will not remain conjectural. And oil or gas beneath the polar ice will not remain there. The motorists want it, and they will pay what they must to get it. In time it will be burned. Yet also in time there will be no more of it to burn. In the 21st century, the last of the reasonably and economically accessible oil will be gone. Ironically, with such an end in sight, the prestigious *Economist of London,* with high prices per barrel and a worldwide scramble for oil underway, has predicted a "rather long-lasting glut of oil" (43). The scramble to get it out of the ground, and the rush to the pump by every driver eager to consume his share, will simply hasten the arrival of the last day. The editor of *Science* viewed this situation and expressed concern in 1975:

> The United States continues to drift toward some form of drastic unpleasantness. Consumption of gasoline exceeds that of a year ago. Domestic reserves and production of oil and natural gas are steadily declining. Total U.S. inventories of oil and its products are below those of a year ago. . . Perhaps the most discouraging feature of the present scene is a failure of U.S. leadership. . . But failure of leadership is not confined to the politicians. What have intellectuals done? What have the universities contributed? Perhaps the worst failure has been that of the mass media (44).

FACTS OF THE MATTER

The raw materials picture is no less discouraging than that of oil. If the energies we have depended on (oil, gas) are running out, the other materials we have relied on are also in increasingly short supply.

The U.S. Bureau of Mines in the figures on the next page has estimated the earth's future raw material prospects (45). The Static Index shows how long known reserves will last under present consumption rates. The Exponential Index shows how long they will last if present growth rates continue. The final column shows how long the materials will last if reserves prove to be five times the current known reserves.

The table shows that a slow growth rate by reducing consumption increases the number of years available before resource exhaustion is reached.

The raw material problem is already apparent in the U.S., where continuing economic health requires a never-slowing pattern of growth. A few years ago, the U.S. needed to import no more than two of the thirteen raw materials vital to industry. Today it must import about half those materials. It is estimated that in less than two decades, nearly all will have to be imported either entirely or partially. In the world resource competition, the U.S.S.R. and China may temporarily be in an advantageous position. Each has vast, undeveloped mineral reserves. But in the U.S., whose industries have been exploiting resources at an accelerating pace for a much longer time, large new findings of key materials are not expected.

Resource	Static Index (Years)	Exponential Index (Years)	Exponential Index Calculated Using 5 x Known Reserves (Years)
Aluminum	100	31	55
Chromium	420	95	154
Coal	2300	111	150
Cobalt	110	60	148
Copper	36	21	48
Gold	11	9	29
Iron	240	93	173
Lead	26	21	64
Manganese	97	46	94
Mercury	13	13	41
Molybdenum	79	34	65
Natural gas	38	22	49
Nickel	150	53	96
Petroleum	31	20	50
Platinum Group	120	47	85
Silver	16	13	42
Tin	17	15	61
Tungsten	40	28	72
Zinc	23	18	50

In an effort to project the effect of various resource strategies, the group that conducted the well-known Club of Rome Project studies has established models for potential courses of future action. The group used computers to show the likely results of particular strategies. The model based on one crucial resource material, copper, and on no recycling, shows a virtual collapse of this material by the year 2200, with large amounts of waste, high pollution, and a small number of products in use. The model demonstrates a considerable improvement when parameters of the copper model are changed to include an extraction tax and a recycling subsidy. Then the model reveals a marked increase in product lifetime, reduction of material per product, and a doubling of recycling.

Models for other materials presumably would show similar results. Whether the models are entirely correct or not, they do indicate important factors in future material usage. Attention to the implications of such models offers a considerable opportunity for more rational future planning. If actively used and continued, model studies would benefit from experience and quite accurate predictions would become possible.

In addition to extraction taxes and recycling subsidies, other approaches to resource conservation should include active programs of reduced consumption. If this factor could be realistically added to the model cited, further improvements would be seen in future prospects for copper. Reducing consumption, of course, enters the area of life styles and human values. The same is true of recycling depletable resources such as copper. Technology can contribute to efficient recycling processes. Technology can help reduce consumption by reducing waste in production methodology. But any realistic and pragmatic recycling program would also require active cooperation on the part of all citizens. They would have to participate by segregating and classifying their waste materials on a regular basis. Establishing such a program would be a challenge to leadership. People would have to be educated to recognize and accept the need for such a program. In this educational process, technologists would no doubt be called on to collaborate with politicians in explaining about shortages and how conscientious recycling can benefit all. Cynics argue that the citizens of the U.S. and other industrial nations are too lax and self-indulgent and spoiled by their economies of plenty to cooperate fully, yet such cooperation would be vital to the success of any program. Cynics perhaps do not take sufficiently into account that people have demonstrated the ability to change both their attitudes and their values when new circumstances arise. If the type of economy that makes wasteful consumption possible, if not a virtue, is replaced by an economy of shortages, the people gradually may with encouragement and education learn the vital importance of recycling and how their own contributions are important. In the U.S., recycled paper has achieved wider and wider use. Some states, such as Oregon, have passed legislation against throwaway containers. Other states have defeated such legislation, but have considered it; and future legislatures will always have the power to reverse the decisions of earlier ones. As time magnifies resource problems, the assumption is that they will do so. The corollary assumption is that citizens must insist on it.

CLEANING UP THE WORLD

Before the 1960s, pollution was a seldom-mentioned concern. Now we acknowledge it as one of mankind's major problems. In Jean-Paul Sartre's play "No Exit," a character defines hell as "other people." If pollution were defined in similar causative fashion, it would have to be defined as "everyone and everything." Only a vacuum could qualify as something that is polluted without polluting. Pollution is other people, but it is also inevitably ourselves. Our automobiles have exhausts. We exhale the same as everyone else. And when we finish with something, we deposit it neatly and pollutingly in the garbage. How can a good citizen pollute by doing his duty, not littering, and putting his refuse carefully in the proper container prescribed by law? Well, something has to be done with that refuse. It has to be carried somewhere and either buried or reprocessed or something. That something more often than not means pollution. The U.S. government, for instance, has ordered New York City and other Eastern cities to stop carrying their refuse out to sea and dropping it in the ocean. New York has been doing so for more than 50 years, and other cities have joined the garbage parade out to the great oceanic dumping ground. But the ocean suffers as a result, and ocean pollution has been carried inexorably toward East Coast beaches. Now the government with its leviathan talent for moving slowly has finally moved. Officials in charge of lakes and rivers around the world have been forced or will be forced to make similar decisions: No more garbage here, please! The dead waters have taken all the death they can hold.

The present is not the past. Our busy, crowded world cannot enjoy the free pleasures of uncomplicated and relatively unpolluted yesterdays. Eighteenth century housewives could throw their refuse out the window for dogs or cats or passers-by to enjoy or dodge as they saw fit. In most of Europe, the good wives shouted "gardyloo" (watch out for the water!), or equivalent, before launching their daily slops upon the air and into the streets. Such romantic times are gone in our modern cities and suburbs. There are simply too many housewives and a woeful excess of refuse.

When populations and industrial output were comparatively small, technology could deposit effluents in streams, lakes, and the atmosphere without greatly damaging the environment or the inhabitants of the neighborhood. However, both populations and industries grow rapidly. The adverse effects of the older methods and manners become apparent. Even shouting "gardyloo" doesn't suffice.

Earth's biosphere (the living environment) is threatened in many ways and from many directions. These threats in turn impinge upon continued technological development. Cleaning up the environment costs not only money, but also in many cases a radical change in habits and values by both individuals and industries. For example, atmospheric decontamination involves, among other things:

1. Removing pollutant gases, such as sulfur oxides, from power plant smokestacks.
2. Regulation of automobile exhausts.
3. Preventing by-products of combustion in steel mills, smelters, and other places from entering the atmosphere.

These three items by no means exhaust the list, yet each poses hard challenges to the public in terms of cost and to technologists charged with effectively implementing each. A few years ago when the idea was first publicized, the necessity of reducing poisonous automobile exhausts was generally accepted. At that time most were against poisoning their neighbors or allowing their neighbors to poison them. But when the apparent costs of preventing automobile exhaust pollution became known, enthusiasm waned. With the immediately available technology, lowering harmful pollutants in exhaust pipes increased the cost of the automobile while lowering its efficiency, and thereby also increased operating costs. Reluctant to pay the price, the average automobile user could almost be seen thinking: "I got along all right before. Why mess things up. Let's go on as before."

The problem is, of course, that the war against pollution confirms the oldest of cliches: You don't get something for nothing. Eve paid for the apple by carrying the responsibility of knowledge. Taxpayers must foot the bill for clean water, and drivers must help pay for cleaner air. Whether U.S. citizens will be willing to pay is the multibillion dollar question. Perhaps the environment is not yet sufficiently polluted in many areas. Smog in the Southern California basin has proved convincing there. Stringent automobile exhaust laws have been established and enforced. Elsewhere, with less conspicuous pollution, it will take longer. The typical U.S. citizen is greatly enamored of his automobile. He will pay whatever he must to keep it running, but may resist paying for good air as long as possible. The present increase in per capita cigarette consumption in the face of known deleterious effects provides no great encouragement to those hoping for rational voluntary action.

In "The Tragedy of the Commons," Garrett Hardin examines the difficulty of convincing individuals to act against their own inclinations for the common good, even though achieving the maximum good is accepted by everyone as a worthy goal. Hardin thinks that the values attached to such variables in life as the "common good" (hard to define but instinctively understood) must be made more specific and more universal. One ghost he sought to exorcise was the popular view that an individual intent on his own gain is thereby serving the public interest (the doctrine among others of Adam Smith, patron saint of mercantilism, small businessmen, and touch-me-not opponents of government interference). Hardin relates the story of the commons, originally published by William Forster Lloyd, an amateur mathematician. Lloyd's commons is a pasture open to herdsmen on a serve-yourself basis. Eventually, the commons reaches the point where its capacity is exceeded by the growing community, and at that point says Hardin the logic of the commons produces "remorseless tragedy." Each herdsman seeks to increase his personal gain by adding to his herd. The animals overgraze the commons as each herdsmen increases his herd without limit in a limited environment. The commons is destroyed, and Hardin concludes with the observation that freedom in a commons can mean ruin for all (46).

Fortunately what sometimes happens in such a commons is that one or more of the herdsmen will recognize the catastrophic course being taken. The herdsmen assemble, arguments are presented, debates are held, an organization springs up, votes are taken, rules instituted, enforcement provided for; and the commons is saved as a local administrative government once more becomes established among a group of men who cannot survive without it.

The tragedy of the commons is easily analogous to the current pollution situation. Factory owners and automobile drivers tend to find it more convenient to do as they please, discharging pollutants into the atmosphere or streams. This free use of the "free" environment works all right for a while—as does the grazing of the commons, until community growth exceeds its capacity. But free use of the environment in this callous fashion inexorably leads to catastrophe as each individual pollutes cheerfully to his own private drummer.

Hardin contends that technology offers few solutions to the problem of the commons. He may be too pessimistic about individuals on one hand and technology on the other. Men quite often have refused to listen to reason, have behaved stupidly, and have, in effect, overgrazed one commons or another. But on occasion they *have* listened to reason. Everyone who can see the consequences of a certain course should offer the arguments of reason. Engineers and scientists, possessing reason as a professional tool, should be in the forefront with such arguments. Technology at the same time must not accept the inevitability of doom once individual selfishness or neglect is moving toward a deplorable end. Technology if nothing else can attempt to prolong the life of the commons . . . or oil reserves . . . or material resources.

Technology can also contribute to the debate on freedom. The concept of freedom is still part of the rhetoric in western political circles, but the concept is seldom defined. It seems to mean that a person should be called free if he can do as he pleases, within reason. The qualification is essential, since "within reason" has imposed a wide range of limitations on western man. The limitations are applied, it might be said, to promote the general welfare, to protect the individual, and to preserve the commons. A traffic light restricts the freedom to cross an intersection at will, but the light affords the freedom of crossing the street safely. Georg Hegel defined freedom as the recognition of necessity. It is necessary to pay taxes so that the commons can be kept up, reseeded, and policed. Paying taxes becomes obligatory and thus an abridgement of freedom, but other freedoms are acquired in the process.

A continuing role for technology is to apply the techniques of reason in helping men recognize and accept necessity. If men and their families are to live without the constant threat of disease from the air they must breathe, clean air must be recognized as necessary. Identifying what is necessary, prudent, and reasonable is a process that can be aided by technologists in their communities wherever they are.

Every commons is worth saving; indeed every commons is essential. Destroying one commons forces herdsmen to try again in the closest grazing land they can find. Thus, a second commons will be threatened. And so on. Prolonging the life of the first commons becomes a conspicuous necessity. By extension, prolonging the life of planet earth is a conspicuous necessity. For all our speculations about the stars, we still know nothing of other grazing lands in our cosmic neighborhood.

MODEL ANALYSIS

What does the future hold? Model analysis by means of large computers provides a new kind of crystal ball. Anything from engineering problems to economic systems to resource management can be studied by running a multiplicity of factors through the computers. The "model," as the structure of factors is called, simulates specific problems or conditions in the real world. Just as pollsters predict how a nation will vote by sampling the attitudes of a few, so the model analyst can predict the future expectations of systems or even countries.

One of the first studies of a large-scale model was considered in *World Dynamics* by Jay W. Forrester, former director of the M.I.T. Digital Computer and a pioneer in model analysis (47). Forrester concludes that industrialization (technology) contains the seeds of its own destruction and presents a greater disturbing force in ecology than overpopulation. He sees collapse resulting from the exponential utilization of raw materials from limited supplies. Forrester thinks the U.S. and Europe are passing through a golden age, with a quality of life that will not be matched in the future. He sees no possibility for underdeveloped countries to achieve the living standards of the industrialized countries, because the "load" on resources will not allow it. He warns underdeveloped countries not to be overly ambitious since in the approaching crash the developed countries will suffer most. The tragedy will be greatest for the people of the developed countries, since they must lose what they have learned to take for granted and to consider essential.

Using Forrester's model, the Club of Rome, headed by Italian industrial manager Dr. Aurelio Peccei, made further studies in 1970 (48). This group ran the model from 1900 to 2100, with resources, population, pollution, food per capita, and industrial output per capita as the main variables. In nearly all cases, the model exhibited a pronounced collapse of all variables on or before the year 2100. The only model run which produced more favorable results in terms of food per capita, less pollution, and reduced use of resources by 2100 involved a case where population and industrial output were severely restricted. This run did not exhibit the large overswing with subsequent collapse shown by most other runs. The conclusions of the group contained blunt admonitions:

1. The result of present growth trends will be sudden and uncontrollable declines within the next 100 years.
2. Such declines may be prevented by control of population, industrial restraints, recycling of materials, pollution controls, agricultural advances, and related efforts.
3. A stabilized system with disciplined effort can evolve. However, the concept of growth as an objective must be abandoned.
4. The sooner controls are installed, the more likely it is that stabilization can be reached. The later controls are installed, the less effective they will be.

The authors of the report do not claim to predict future events. They simply offer the results of their model for discussion. Whatever inaccuracies the model contains, it seems clear that continued exponential growth in a finite universe will lead to decline, and that stabilization must be the alternative to untimely exhaustion.

In 1976, American political candidates were warning the American people realistically for the first time of resource limitations. Americans heard that they must learn to expect less, that economic growth is not an unlimited process. These were startlingly new sounds on the American political scene, and there was something even more startling: many Americans were listening attentively. Perhaps in 1976, the time had finally arrived for people to hear and understand what U Thant, Secretary-General of the United Nations, had said in 1969:

> I do not wish to seem overdramatic, but I can only conclude from the information that is available to me as Secretary-General, that the Members of the United Nations have perhaps ten years left in which to subordinate their ancient quarrels and launch a global partnership to curb the arms race, to improve the human environment, to defuse the population explosion, and to supply the required momentum to development efforts. If such a global partnership is not forged within the next decade, then I very much fear that the problems I have mentioned will have reached such staggering proportions that they will be beyond our capacity to control.

These warnings become more relevant in the light of prevailing trends at the start of the last quarter of the twentieth century. The conclusions of model analysts also tend to be gloomy concerning the future prospects of the human race and technology. If we do not heed the warnings, the worst predictions may come true. And perhaps the warnings will not be heeded. But gloom notwithstanding, helpful choices are still available. In energy, for example, good prospects exist if we possess the will and patience to conserve what we have while alternate energies are developed. New technology in resource use, energy conservation, and material recycling doesn't guarantee Utopia, but new technology does support the hope that mankind can, with sufficient and consistent effort, avoid losing control. Men still have the capacity and the tools to save themselves.

Perhaps as an antidote or at least a counterbalance to projections of failure, we should quote a technological optimist. Addressing scientists and engineers at Auburn University in 1969, a modern enthusiast for technology, Buckminster Fuller advised against overspecialization in science or life, and paid this tribute to man:

> We've all been working under the assumption that man is destined to be a failure. I say man is quite clearly like the hydrogen atom: designed to be a success. He is a fantastic piece of design; it is completely wrong to think he is meant to be a failure. I assume he is supposed to be a success and that he is supposed to use his mind to make himself a success (49).

Fuller added something he had originally said in 1927: "I'm going to commit the rest of my life to exploring the whole matter of doing more with less." Doing more with less is precisely where the challenge to technology reaches us today.

REFERENCES

41. Rose, David J., "Energy Policy in the U.S.," *Scientific American*, Vol. 230, No. 1, January, 1974, pp. 20-30.

42. *The New York Times*, "Secret Accord on Minerals in Antarctica," July 18, 1976, p. E7.

43. Yergin, Daniel, "The Economic Political Military Situation," *The New York Times Magazine*, February 16, 1975, p. 10.

44. Abelson, Philip H., "Absence of U.S. Energy Leadership," *Science*, Vol. 189, No. 4196, July 4, 1975, p. 11.

45. Mineral Facts and Problems, 1970, U.S. Bureau of Mines.

46. Hardin, Garrett, "The Tragedy of the Commons," *Science*, Vol. 162, December, 1968, pp. 1243-1248.

47. Forrester, Jay W., *World Dynamics*, Wright-Allen Press, Inc., Cambridge, Massachusetts, 1971.

48. Meadows, D. H., Meadows, D. L., Randers, J., and Behrens, W. W., *The Limits to Growth*, Universe Books, New York, 1972.

49. Fuller, R. Buckminster, *et al.*, *Approaching the Benign Environment*, Collier Books, New York, 1970, p. 96.

QUESTIONS FOR REFLECTION

1. If you were "Dictator for a Day" what would you do? For a year?

2. How do you react to the following excerpt from Taylor Littleton's preface to *Approaching the Benign Environment*, from the Franklin Lectures at Auburn University, 1969: "All three speakers, each in his own special way, expressed deep concern about the narrow outlook which contemporary education tends to foster in our engineers and scientists, who, paradoxically, are being called upon increasingly as advisors and decision makers to help society resolve many of the social and economic problems which the new technology has created."

3. Would an economy directed by technologists be better than the one we have? Would it be different at all?

4. Weigh the proposition: "These are the best of times for the human race." Which view do you prefer: "Yes" or "No"? Why? Are there potentially better times ahead?

5. Do armaments promote peace? Does the atomic bomb make war less likely because of the "balance of terror?"

6. When the U.S. rations energy, which industries or groups should be favored?

7. Assuming effective model studies are possible, what, if anything, can or should be done with the predictions that result?

8. Is the United States currently experiencing its best era with conditions destined to worsen steadily from now on?

9. Viewing earth as the commons discussed by Garrett Hardin, can the tragedy he foresaw be avoided? How? Is the tragedy *likely* to be avoided?

10. What is the human race doing to assure failure? What should it do to enhance the prospects for success?

6. Technological Solutions

"The road backward from our day to some earlier 'simpler' time when everyone grew his own vegetables and rode a bicycle to work is as uncharted and as filled with incalculable suffering as any of the ways forward. In fact, such a regression, attractive as it might seem to a 'nostalgic' generation is nothing more than a dream."

"The Talk of the Town"
***The New Yorker*, October 21, 1974**

THE PERILS OF SOOTHSAYING

Improvement in man's ability to predict the future and to plan ahead encourages progress. Concomitantly, prophetic gloom about the future leads to impotency. As Pope insisted in his *Essay on Man,* "Hope springs eternal in the human breast." The author of *Proverbs* did not refute this, but noted that "hope deferred maketh the heart sick." To take away expectation of good or to prophesy doom serves to baffle the will and stifle the spirit.

Would *Homo habilis* have managed to continue his struggles if he had foreseen all the trouble ahead? The answer is probably yes. He would have hoped that by knowing the trouble he could do something to lessen it. He would have hoped for the best and settled for something less. Because of the inveterate impulse and wish to hope for the best, life produces more optimists than pessimists, more yeasayers than doomsayers.

In preceding chapters, human hunger, material shortages, rumors of war, and energy crises, led to dire predictions and warnings, by means of model analysis and other ways of looking through a glass darkly. Fortunately, the human race reasonably often demonstrates the predilection of soothsayers to end up with their feet in their mouths. It is not that the soothsayers add up their variables incorrectly or abandon reason for moonshine in forming their predictions. An individual or group simply does something remarkable that tosses all the current crop of predictions into a cocked hat.

In the historical panorama, technology has been a frequent ally of man in frustrating predictions. In fact, that irrepressible pair, a curious mind and busy hands, have been so successful in upsetting prophetic applecarts, knowledgeable persons become cautious about making new predictions.

In the case of nearly every important historical development, the reigning soothsayers of each age ridiculed it as impossible. The world was flat, they proclaimed; then Columbus sailed West to a new world, and Magellan circled the earth. The grandfathers of the "flat world" advocates insisted that the earth is the geometric center of the universe, but the facts wouldn't support such a fantasy. Copernicus, Kepler, and Galileo began the long process of determining the truth. Airplanes wouldn't fly, the experts predicted; Langley and the Wrights proved them wrong. So they agreed man can fly; but he can never fly out of this world, declared the doubters. Then Neil Armstrong and Buzz Aldrin walked on the moon.

Technology and curious minds have an impressive record of discovering new truths and methods that refute the warnings and predictions of doomsayers, standpatters, and those of the soothsaying profession who deal in negatives. This fact, of course, does not justify another fantasy—that science can heal all ills and solve all problems. Technology alone has never accomplished miracles. It has provided tools with which men can get the job done, but men have to do it. They are the ones who use the tools and find some way to muddle through. The assumption that science will end pollution, the energy crisis, hunger, and war, while mankind naps on the front porch is naive. It is also a dangerous invitation to complacency. Men, not their tools, must face facts and make the hard decisions. They must resist the temptation to expect miracles from technology. The very success of technology during the past 200 years leads to complacent assumptions. If man can reach the moon and Mars, he can eliminate hunger, pollution, overcrowding, crime, war—can't he? If organized technology delivers the atomic bomb, elaborate computer systems, industrial complexes, and nuclear energy, then curing cancer or heart disease or mental despair should be child's play for clever science. Right?

Such assumptions compliment the past achievements of technology. They say nothing and know nothing about future achievements. It would clearly be naive and dangerous to expect too much from future technology. Scientific knowledge has always been achieved by moving forward in the darkness and waiting for occasional brilliant flashes of light. The comment has been made, quite seriously, that technology has been lavished on mammoth athletic stunts such as walking on the moon, because the common cold is too much to tackle. Technology has carried out impressive organizational efforts and elaborate planning while combining intricate technologies to make an atomic bomb or to probe space. But disease still kills. The weather still cannot be predicted with reliable accuracy. And for all his technological strides, man seems closer to the possibility of extinction today than ever before in his history. Thus, *The New York Times* in an exuberant editorial praising scientists for the

Viking landing on Mars, logically included this exceedingly ominous bit of optimism:

> Barring the catastrophe of a nuclear war, it is not stretching the imagination too far to suppose that by the time of this nation's Tricentennial, Mars could be sustaining a human population (50).

Barring the catastrophe of a nuclear war, one wonders if by the time of the U.S. Tricentennial, earth will still have a billion and a half hungry inhabitants, or will the number be five billion, with more in Kansas City, Fort Worth, and Oshkosh. Based on recent trends, it would seem a better bet that tiny, expensive outposts of humanity will exist on other planets than that problems closer to home will have been solved.

A tool, not a panacea, technology does what men drive it to do. So far, men apparently have preferred majestic stunts in space to eradicating pollution in the neighborhood. That is no doubt true for a very human reason—it is clearly easier to visit Mars than to solve the grimly persistent human problems next door, or indoors.

LIMITATIONS OF TECHNOLOGY

Technology cannot deliver the answers to the tough problems. It is the human mind and will that ultimately must deliver every answer. Lecomte Du Noüy in *Human Destiny* (51) wrote hopefully of man's future, but the hope contains a catch:

> Let every man remember that the destiny of mankind is incomparable and that it depends greatly on his will to collaborate in the transcendent task.

The Individual Will to Collaborate

"Well, there's a meeting this morning. We have bridge tomorrow night. Golf Wednesday. Golf Thursday. I've forgotten what it is, but something Friday. I hate to break into the weekend, but listen, come over Sunday afternoon. We'll have a drink and collaborate for an hour or so." Good luck to us all.

If the problem is ourselves, the answers cannot be in technology. In Du Noüy's view "humanity has not reached the age of reason and its efforts are still on the scale of the tribe." He wrote of a goal that he considered higher, and harder to reach than the sort of cleverness with apparatus that could perform prodigious feats such as space exploration:

> Actually the progress of science is measured by its practical applications and not by the evolution of philosophical thought which results therefrom; and yet the last is more important than the first; it is, or should be, the real goal of science (52).

Expectations in much of this century have been that technological innovation can answer all questions, heal all wounds, liberate mankind for the good life, and bring on the millennium. These expectations were fostered and maintained by a widespread and deliberate misunderstanding of what technology is all about. One of the wisest and kindest of science critics, Joseph Wood Krutch, observed that "science, though it fulfills the details of its promises, does not in any ultimate sense solve our problems." Krutch continued:

> We went to science in search of light, not merely upon the nature of matter, but upon the nature of man as well, and though that which we have received may be light of a sort, it is not adapted to our eyes and is not anything by which we can see (53).

And he concluded:

> We are disillusioned with the laboratory, not because we have lost faith in the truth of its findings but because we have lost faith in the power of those findings to help us as generally as we had once hoped they might help . . . if we were compelled to sum up our criticism of modern science in a single phrase we could hardly find one better than this last—that it does not seem, so surely as once it did, to be helping us very rapidly along the road we wish to travel. We cannot make physical speed an end to be pursued very long after we have discovered that it does not get us anywhere, and neither can we long devote ourselves whole-heartedly to science except in those departments—like medicine, for example—that accomplish not merely results, but results which have an ultimate value (54).

There's that word again, value. And it is easy for technologists to sympathize with the point Krutch tries to make. Technology demonstrated unquestionable excellence in keeping its promises—putting a lander on Mars, for example. But the value of those promises is a broad human question with various and inexact conclusions. Philosophers such as Krutch warn that disappointment will haunt those who expect everything from science, because science cannot deliver. It can solve problems. Properly directed by the human mind, it can throw floodlights on the plain when values are being identified. Yet science must take its lead from man, rather than lead man.

To derive the optimum benefit from technology, it must be asked to make and keep the best promises. What those promises should be, however, lies beyond technology, and appropriate direction comes from elsewhere. In the end it may come from those who sign the checks, and in democracies such as the U.S., the money usually is spent on projects approved by the majority of citizens. Technology may be told to build the greatest war machine the world has ever known, to construct the most elaborate network of highways, to end the pollution of lakes and rivers, to stop a crippling disease such as polio, or... The list is endless, and the priorities are established not by technology but by citizens as they see or are led to see their needs. A remarkable thing about technology is its nearly perfect batting average when sent up at the top of the seventh inning to achieve a certain goal. Therein we find the optimism of modern science; that it can do what it is asked and properly financed to accomplish. But setting the goal—there's the rub!

Only frustration and balked desires are harvested when we expect technology to surpass its own limitations. For one thing, it cannot step inside the human mind with a mop and scrub bucket and clean up the mess there. That is not a task within the purview of technology. Mind-cleaning is an individual challenge with only human values for soap.

Technology aids immeasurably, but not outside the limits of its own nature. Automation in instruments and machinery does not carry over into the human mind. Mind selects the problems. Mind determines the shape and direction of technology. The robot on Mars does only what it is told. The telling of the robot is something other than technology. Mind pulls the strings so that technology can go into its act. Human values also grow, like flowers or weeds, in the mind, and they affect decisions about which strings should be pulled and when.

Scientists often play hunches. They develop successful theories by remarkable leaps in the dark, followed by failures, slow, tedious observations, and then further leaps in the dark. A fascinating illustration of this process is found in *The Double Helix,* James Watson's account of the work done by himself and Francis Crick at Cambridge, leading first to a determination of DNA structure and then to the Nobel Prize. In the work at Cambridge, Watson was the "leaper" and Crick the meticulous, step-by-step worker. Together the two men accomplished brilliant science (55).

Often genius has difficulty explaining just how it arrives at solutions. Sometimes it seems close to an accident as when Watson's mind leaped to the idea of the double helix as the structural component of DNA. But Pasteur's reminder that "chance favors the prepared mind" should not be forgotten. Preparation and perspiration are both required in ample measure before "chance" has a chance. Ultimately, though, the great breakthrough may seem a product of inspiration or a leap in the dark, and neither inspiration nor leaps in the dark as yet have been programmed for computers. Nor has a formula for either been found. This is one limitation of technology that both exponents and critics need to know. Technology is an arm of human effort and decision; it moves only when commanded by the human brain.

The evidence shows that technology when intelligently commanded has strengths and perhaps even solutions to offer in many areas of human concern.

ARMS CONTROL

It is a question whether the U.S., Russia, or any other powerful nation has ever seriously sought genuine arms control. Investments in weapon technology have continued to be enormous, and finding ingenious ways to kill has received more energy inputs than finding ingenious ways to peace. However, if there were a true will to peace and a sincere effort, technology could help ensure it. For example, underground nuclear explosions were not included in the original test ban treaty because some argued that these could not be distinguished from earthquakes. Advances in knowledge of seismology and in the placement of detection instruments have removed that limitation, and technology now makes reliable detection possible. Nevertheless, there has been no rush to implement a ban against underground testing. Some countries (i.e., France and China) have even continued exploding atomic weapons in the atmosphere, despite known hazards from radioactive fallout, and recent evidence that nuclear explosions may endanger the protective ozone layer in earth's ionosphere (56).

Another proven means of monitoring weapon activity anywhere on earth is available: satellites in orbit, capable of examining installations of all kinds quite minutely, to photograph them and transmit the picture to earth via radio. Even camouflaged and underwater devices can be located by infrared or other types of detectors.

Clearly, technological means are available to police and monitor any peace that the various nations will maintain. If peace is not maintained, it will be the will to peace that fails. Human values that choose fear instead of trust, and war instead of peace will be the adversaries, not technology. To assure the choice of peace and the use of surveillance technology to keep the peace, technologists have a responsibility to involve themselves in the politics and debates of peace. They must explain what can be done and how far it can be trusted.

THE COMING ENERGIES

With traditional energies either in short supply (petroleum), threatened by depletion (gas), troubled by politics and unsolved technical problems (nuclear), man's energy future has become a technological challenge. Fortunately, a wide range of alternate energies are ready for development. The race is on among these alternate energies, and essentially it is a race of technologies. Every runner in the race should be encouraged. It seems apparent that mankind will be able to utilize all the energy made available to fuel civilization. Keeping energy conveniently and reliably available is a function of technology, and progress on a number of energy fronts continues.

Hydroelectric

Water power resources are already substantially developed in the U.S. Proposals to increase energy from water power have led to big dam projects and other developments seriously questioned by environmentalists. In the U.S., substantial additional energy from water power is unlikely. Other countries, such as Russia and China, still have sizable water resource possibilities.

Geothermal

Geothermal energy has been successfully used for many years in various parts of the world (e.g., Italy, California). New developments are underway (e.g., the Philippines) or projected in those areas where use of geothermal energy is practical. The energy is acquired by drilling wells in areas where underground heat is available close to the surface. Steam or hot liquids when conducted to the surface operate conventional electric turbines. Geothermal energy will not be a large-scale answer to energy shortages, but can relieve pressures on conventional energies in many areas. The Imperial Valley of southern California and northern Mexico is one region with molten magma suitably close to the surface for geothermal energy use, and in the U.S., ERDA (Energy Research Development Administration) has sponsored test facilities at East Mesa in the Imperial Valley. New technology is also envisioned that will make deep wells practicable for tapping the constant heat energy available in the earth's interior. The same as most energy sources, geothermal energy developments will have environmental side effects, but these may be less than the pollution effects of oil and the hazards of nuclear energy. Development of geothermal energy is an active technological undertaking that should in time pay important energy dividends for many regions.

Wind Power

This is a form of solar energy available in various locations. Technologists are working to capture this energy and store it for use. In the mid-1970s, a number of large-scale wind power development projects were in progress. ERDA was sponsoring a series of 100-kw wind generators to prove the practicability of wind energy in diverse areas. The hope to use wind energy in conjunction with solar energy had become an expectation as technology coped with some of the technical problems. The environmental effects of wind energy development were expected to be insignificant aside from possible aesthetic objections to the presence of giant wind machines in large numbers. Technological emphasis in the 1970s was on the reduction of wind energy costs and on energy storage. Since wind occurs intermittently, storage is the key to its success as an energy source. Thus using wind as a supplementary energy together with solar energy is receiving careful attention. In 1976, technology for economical use of wind energy still had a long distance to travel. For instance, in the July 25, 1976 issue of *The New York Times* it was reported that New Jersey had tabled a plan to erect 300 windmills along its coastline for generation of electricity. A spokesman said they planned to wait until the "technology becomes more feasible." Expectations of that occurring are optimistic.

Nuclear Fission

U.S. energy policies in recent years have tended to put much hope and money into this energy form that was supposed to answer all mankind's requirements after the end of the oil. The uranium supply appears limited to about 20 years, but the breeder reactor theoretically can be available before that time, indefinitely extending nuclear fuel supplies. The U.S. currently spends nearly half a billion dollars per year on fission power, much of it going into research on the liquid sodium-cooled version of the breeder reactor.

In the 1970s, nuclear fission has become intensely controversial. A series of minor mishaps or failures at various nuclear installations in the U.S. have dramatized the question of public safety and the danger of "melt down." The disposal of waste also remains a critical, unsolved problem. One of the wastes, plutonium, has a half-life of thousands of years, and it has been estimated that one ounce could form 10 trillion particles of plutonium-dioxide and set off an international lung cancer epidemic. The U.S. nuclear industry in 1974 produced 8,000 pounds of plutonium. When nuclear expansion reaches projected levels, up to 600,000 pounds of plutonium could be produced annually. Disposing of such dangerous by-products confronts the nuclear industry with a complex technological challenge. Burial has proved a faulty method because of leakage. Disposal schemes range from burial in deep salt chambers along the U.S. Gulf Coast to being dropped on the two-mile-deep ice of the Antarctic.

In addition to the problems of waste materials, recurrent troubles have been experienced by many of the nuclear reactors now operating in the U.S. and elsewhere. Much doubt has been cast on the future prospects of nuclear fission. Original high expectations have been eroded by breakdowns and safety concerns. The future for nuclear fission consequently is in doubt and represents an area of energy technology in which concentrated effort still continues. The goals of safety plus efficiency are still to be reached, but many experts such as physicist Hans Bethe believe they are reachable, and that mankind has no choice but to continue trying.

Nuclear Fusion

The scientific feasibility of nuclear fusion has not yet been demonstrated, but feasibility may be proven by the early 1980s. The technology required to accomplish atomic fusion is predicted by 2000, and commercial energy from fusion by 2025. Deuterium, plentifully available in seawater, is mentioned as the ultimate fuel in an ideal nuclear fusion cycle.

Fusion will have marked advantages over nuclear fission. It will be a relatively "clean" energy without dangerous wastes. The threats to public safety associated with nuclear fission will not be characteristic of fusion systems. Thus, fusion seems to offer a long-term answer to man's energy quest. For this reason, research on an international scale currently displays a spirit of cooperation. Fusion, based on a Deuterium fuel cycle, would benefit all mankind by eliminating energy concerns for thousands of years. Proving that it is possible and developing the technology necessary to exploit that possibility is where fusion research is now.

Temperatures approaching 100 million degrees are required. These are considered achievable temperature levels, and the difficulty of sustaining a fusion reaction for several seconds is also considered close to resolution. Some experts contend it is only a matter of time until the attainment of fusion energy. The experience with nuclear fission, however, has sobered those who thought nuclear energy provided an easy answer to man's energy appetite. The nuclear feast may be a long time coming.

Solar Energy

The solar radiation reaching the earth contains approximately 30,000 times the total power used by man. Taking direct benefit of this energy is an enthusiastic energy crusade that began with special eagerness and urgency in the 1970s. The use of large solar collectors on solar farms of many acres illustrates one approach to utilization. The technology for receiving, storing, and transmitting solar energy is still in the developmental stages. Practical solar heating and cooling systems have been designed and installed in family homes. This special application seems the most promising means of harvesting solar energy efficiently. Relatively small solar collectors can heat a well-designed and insulated house or building. Deriving large-scale energy for cities is technologically much more complicated. A solar panel 15 miles square would be required to power New York City (57). This clearly is not economically feasible. Another dilemma of solar energy technology is storage of power for use during dark periods. Wind power or other complementary energies is one approach to the problem. Power-generating satellites, and even orbiting solar energy communities in space are other quite serious plans put forward theoretically for pragmatic implementation. Except for residence or building heating, solar power for the present must be classed with nuclear fusion, wind power, and geothermal energy as an "exotic" possibility.

Hydrogen

The use of hydrogen as an energy fuel is being investigated by a number of engineers. Hydrogen offers certain advantages. Burning hydrogen produces no impurities. It could be transmitted through existing gas pipelines, bottled for use in vehicles, and stored in reservoirs. Hydrogen would have to be made from other energy sources, preferably non-fossil and plentiful. That is why at this stage, hydrogen energy is still an exciting idea rather than a commanding reality. Technology must resolve the initial challenge of economically converting some non-fossil fuel source, such as solar energy, to hydrogen.

Alcohol

This heading includes fuels such as ethanol, methanol, methane, and others produced from vegetation. These fuels offer hopeful new energy sources. Considerable energy could be obtained by utilizing waste vegetation. More might be generated by deliberate cultivation of plants. "Energy farms" are practical possibilities in many areas. Farm wastes, including manure, can produce methane, which offers many of the benefits associated with hydrogen. These fuels are practical and the technology for utilization is straightforward. Active efforts are underway to supplement traditional energies with these new ones. In Brazil, conversion of the cassava plant to ethyl alcohol is considered a long-range answer to Brazil's energy needs. In the U.S., manure gasification plants have been constructed to provide biogas, which can be carried to American homes via natural gas pipelines. Conversion of peat deposits to methane is a project under consideration for Northern Minnesota. Harvesting seaweed or kelp off the California coast for processing at inland plants to produce methane, as well as food and fertilizer products, is a project funded by the U.S. Navy and the National Science Foundation. Many cities are committed to the plan of utilizing urban wastes for the production of energy in one form or another. Technology advances on many fronts to make these practical fuels available on a significant scale, and indications are that success is within reach.

Coal

The U.S. has about 1,600 billion tons of coal in reserve, greatly surpassing the world's proved reserves of crude oil. America's coal is sufficient to last several hundred years at the present rate of depletion. U.S. governmental policies have been restressing the use of coal, following a long period of decline in coal consumption. Since the beginning of the twentieth century, coal had been methodically shoved aside by oil. The future use of coal on a much larger scale than in the recent past seems inevitable. Coal will be employed directly as a fuel in power plants, and it will also be converted to oil or gas. Coal, sometimes called "old filthy," has familiar drawbacks: Its removal is difficult. Coal miner deaths and illnesses are among the social costs. Strip mining often ruins vast regions of the earth, and most of the western coals containing small amounts of sulfur must be strip-mined. Newer techniques may facilitate reclamation of strip-mined lands, but these techniques depend on water, which can be scarce or unavailable for such purposes in the arid west. Direct burning of coal produces pollution, with sometimes dangerous materials (e.g., sulfur, nitric oxides) delivered to the atmosphere. Sulfur removal has received attention from many researchers, but as yet an effective, economically feasible technique has not been achieved. Technologists also point out the shortsightedness involved in burning coal as fuel. Coal has greater value, they contend, in supplying hydrocarbons for chemical raw materials. Developing and using other energies to save coal for the future is strongly urged. The energy facts of life, however, suggest that millions of tons of coal will be burned before such alternate energies become sufficiently available.

Oil

With 80 percent of American households dependent on the automobile in some fashion, the demand for oil will remain high as long as it is available in the vast quantities required at prices that are not prohibitive. Approximately half of U.S. energy needs were supplied by oil in 1976, contrasting with 14 percent in 1920. Despite new fields in Alaska, offshore drilling, and other measures, it is unlikely that U.S. and Canadian oil can do better than hold at present levels (58). Because of the rising need for oil, increased imports are the indelible handwriting on the wall. Efforts at oil conservation in the U.S. have largely proved to be nothing more than rhetorical gestures. The challenge to technology in connection with oil is finding and perfecting alternative energies. Another may be to participate in an essential educational process to instruct consumers that oil is not a commodity infinitely available. Citizens need to learn that conservation and energy frugality are in their own best interests. Those carrying the bona fides of science may alone be capable of communicating this truth.

Gas

As natural gas supplies dwindle, it may be used less in industry and electrical power production, and reserved for home heating. Gas is expected to furnish approximately 15 percent of the total energy requirements in 1985, slightly more than coal. Natural gas in a sense has been a political frisbee between producers and government regulators. Interstate prices have been fixed by the federal government in the U.S., and producers insist these prices are the main reason new natural gas supplies are not being aggressively sought and developed. Whether gas suppliers exert a subtle form of business blackmail for higher prices is, in practical terms, irrelevant. Gas shortages do exist and severely in some areas. With higher prices, off-shore gas deposits would be more attractive investments for development. The same is true of the project to extract methane from coal mines. Methane explosions have been perhaps the worst dangers confronting coal miners. Venting off methane would reduce this danger and simultaneously supply natural gas to relieve shortages. Vast quantities of natural gas are also thought available in the Devonian shale of the western U.S. The USGS estimates that there may be up to 14 trillion cubic meters of natural gas in this shale (59).

Oil Shales and Tars

Tremendous deposits of oil shale await developers in western Wyoming, Colorado, and eastern Utah. With more easily accessible oil supplies hitherto available, this shale was not the object of intense speculation and planning as it now is. To produce 25 gallons of oil, at least one ton of shale must be mined. Present methods require considerable water, which is scarce in the region where most of the oil shale is located. It now appears that shale at best can furnish only about 10 percent of U.S. oil needs; nevertheless, this 10 percent will be welcome. Oil sands also occur in the U.S. and Canada. Oil in these may be easier to recover than in shale, however processing costs remain high. Here too technological developments, facilitating oil recovery, could change a minor energy source quickly to a major source.

Biowaste

The Union Electric Company of St. Louis, among several such organizations, routinely uses city wastes in a power plant. The energy potential of waste products presumably will be explored by all cities and communities with a need for extra energy. In addition to energy, putting waste matter to use in this fashion will alleviate the disposal problem and reduce pollution. Energy from biowastes represents a significant turning aside from the past. The old wastefulness, at least of energy, may be enjoying its final performances. Extracting energy, even from garbage, is revealing about the future, when every morsel of energy will be tracked down and put to use.

THE LOGIC OF CONSERVATION

After generations of encouragement from corporations, governments, and constant loud advertising to "use more," Americans now and in the future will increasingly become aware that the new morality, compelled by the energy facts of life, is "use less."

The energy glutton is notified by "no gas" signs, painfully higher prices, and polyphonic warnings with funereal organ music in the background that there simply isn't sufficient energy left in the cupboard to go on as before.

In 1976, a writer in *The New York Times* called conservation the "least expensive, most reliable, safest, and least polluting source of energy we can tap" (60). More than half the energy consumed in the U.S. has been attributed to waste (61). Virtually all energy programs now stress energy conservation as a vital need paralleling the development of energy sources. It is recognized as axiomatic that forceful conservation can delay the energy crisis perhaps as much as two decades, thus allowing time for development of conventional energy sources as well as alternate energies. Many knowledgeable observers however believe that we are moving at a snail's pace when we should be hustling like a greyhound. Dixy Lee Ray, former head of the Atomic Energy Commission, said, "I can't agree with an energy conservation policy simply founded on making it cost more. We ought to be conserving on a scale that nobody's even talked about yet, except for a little flurry during the embargo. And time is wasting" (62).

Philip Abelson suggested that America needs a "Conservation Ethic," which he defines as an "outlook that gives legitimacy and even virtue to saving energy rather than wasting it" (63).

Given the necessity of conservation, the question becomes: How can the glutton cut down? Where can energy be conserved and how?

Transportation

Man's love affair with his very own internal combustion inefficiency machine (i.e., automobile) needs to begin tapering off. Transportation requires more than 25 percent of America's total energy budget, second only to industrial consumption of 40 percent. Oil or oil derivatives constitute the main sources of this transportation energy, which is consumed by planes, trucks, and more than 100 million automobiles used daily in the U.S.

Conservation in this area does not represent a complicated challenge. It is really quite obvious, and a number of directions offer significant energy savings.

More efficient use of energy would see trucks and planes increasingly replaced by trains, city buses replaced by subways or interurban transit systems, and most important, automobiles replaced by public transportation. To implement efficient public transport or a successful national railroad system requires extensive public financing and cooperation. This contrasts with the present situation in which transportation costs are largely private. The typical American citizen spends $1,500 a year on an automobile, with minimum complaint. He would no doubt protest bitterly if taxed even half this amount for public transportation. The fact that he pays steep gasoline taxes for highway construction does not escape his attention, but it does escape his wrath. He persistently considers the currently suggested alternatives to the automobile unacceptable.

The automobile is the most inefficient transportation device currently used. In America, the automobile rarely carries more than one passenger. Up to two tons of equipment are utilized to move one person. To accommodate the driver and his automobile, investments in highways and parking places continue to rise astronomically. Installation of bicycle lanes, rapid public transport of all types, and logical discouragement of automobile travel seem mandatory, among similar steps, in the period ahead.

Technologists can contribute both in an educational as well as a technical sense. Technically, scientists and engineers will try to solve the challenge of obtaining maximum benefit from available energies. Smaller, energy-efficient automobiles, for instance, are an immediate and urgent need. Equally urgent will be the task of persuading spoiled Americans to use them, and in the task of persuasion, technologists should and must begin having a voice. Part of the problem in the past has been that the public heard only marketing and sales people, while technologists aware of dangerous choices being made held their counsel. The time is past when technologists should keep quiet. Their input is indispensable to persuade the public that there are alternate, less expensive, less dangerous, and more comfortable methods of getting to work than the ubiquitous flivver. Technology will perform the research necessary to perfect new methods of transport. Among the exotic possibilities are levitated trains, turbine-driven engines, computer-controlled traffic signal systems, and transport-efficient shopping centers and cities. Technologists will also find it useful, indeed, essential, to prepare the public for cooperative acceptance.

Heating and Cooling

Better building practices with particular attention paid to insulation can save much energy. Both commercial buildings and residences can benefit from serious efforts to conserve energy by preventing heat losses. It may be that growing populations and decreasing energy supplies will make the one-family housing unit obsolete, since like the automobile, it is not energy-efficient.

Government research funds are now being invested (finally, say critics of government slowness in the energy field) to perfect solar heating and storage equipment for buildings and residences. Progress has been made that encourages the hope for use of solar energy on a large scale in this application.

Air conditioning consumes a prodigious quantity of energy, and the need for more efficient and better insulated devices is obvious. It may be necessary in an energy-poor society to limit or reduce air conditioning. Alleviation of air pollution would make this easier.

Tax incentives and low-interest loan programs have been recommended as means of promoting better insulation of buildings as well as structural designs contributing to energy efficiency.

Industries too have been actively cooperating to reduce energy consumption. In March, 1976, *Chemical Week* reported on a number of industries that had repaired "steam leaks" and embarked on vigorous savings programs. The magazine noted: "The Federal Energy Administration apparently will be counting heavily on industrial energy savings to help meet its revised energy independence goals" (64).

It is expected that technology in time will provide heat storage devices to save heat energy now being lost from buildings and residences. Innovative thinking and effort by technologists in connection with both houses and offices should pay important energy-saving dividends.

Electrical Energy

Here again we badly need conservation. Electrical energy generation and transmission constitute one of the more inefficient energy systems, although the efficiency surpasses that of the automobile. Only about 35 percent of the energy in fuel used for electrical generation reaches the consumer. The industry has a good record of improving power plant efficiency, but much loss is unavoidable. Substantial research in electrical power generation and transmission should reduce both energy losses and costs. Direct electrical generation, such as use of the Magneto-Hydrodynamic (MHD) generator, and use of cyrogenics in generators and power cables are among developments showing promise. Further technological effort can produce worthwhile results. A low cost, practical electrical energy storage device could reduce costs by more than a third. Other than pumping water into an elevated reservoir with "off peak" power and releasing it for on-peak generation, no large-scale energy storage system has yet been invented.

A non-technical approach to electrical conservation involves regulatory and tax policies adjusting the present electrical rate systems to discourage large usage, rather than to encourage such usage as the current systems often do. The contributions of technologists in this social/political move would be valuable.

Several areas in the U.S. have begun an altered rate system with some success. During the winter of 1975-1976, Vermont used staggered utility rates, charging more during heavy-use periods. Signal lights were installed in homes to indicate when higher rates were in effect, and this proved an effective way to conserve energy. Vermonters changed their living habits, eating and sleeping at different hours, to take advantage of the lower rates. This effectively lessened pressure on electrical utilities during peak hours.

Inevitably, electrical energy use will continue to grow because of its convenience and utility. It may be, however, that rationing of electricity will become necessary at some time in the future, and that obligatory rather than voluntary cooperation as in the case of Vermont will also be required.

Recycling

As indicated earlier, several key minerals such as copper and tin are in seriously short supply, and might run out within a few decades. To alleviate continuing depletion of vital minerals, recycling, an infant but growing industry, offers much promise. Many new schemes are being devised to recover materials from garbage. If successful, it may become practical to mine old waste dumps. Ferromagnetic materials are easily recovered with electromagnets. Non-ferrous metals can be separated by grinding up the waste and using centrifuges or air blasts that operate on density variations to complete extraction.

By offering economic incentives, aluminum salvage efforts have enjoyed some success. Bottles are salvaged in Oregon by legally mandating deposits. Other states have considered similar legislation. A referendum on the subject was held in various states during the 1976 elections, and despite some electoral setbacks, proponents seemed determined to keep trying until the general public fully understands the need.

To recover useful minerals from refuse more effectively, much better results would be obtained at less expense if better separation of refuse items could be made by the consumer at the source. Obtaining this social cooperation is probably impossible on a voluntary basis, since bad habits tend to hold firm until effective pressure is brought to bear. Legislation combined with informative communications efforts, however, could serve to break old habits and establish new ones. Technologists should be involved in any such communications efforts, to add technical credibility.

WILL MAN'S FUTURE BE SAVED?

Technology can contribute indispensably to the solution of man's technological problems, but it cannot do everything. Man himself inextricably is involved in both the problems and the solutions.

Political decisions will be required concerning pollution, transportation, waste materials, population, energy, automobiles, and a multitude of others. Technologists can and must communicate the true facts of our condition and our prospects. Technologists also can argue to the best of their ability for a course of action that seems to them superior to another. Finally, though, men will decide through their local and national organizations which choices to make. From a technological vantage point, the future looks bleak for mankind without united action on many fronts to reach essential goals. This action must be taken with full cognizance of and attention to the earth's increasingly known limitations.

Many of the decisions ahead will not be easy for Americans because they involve hardships, belt-tightening, a change from a consumption to a conservation attitude. Strength of will and intelligent determination are going to be obligatory, and some observers wonder if the citizens of the endowed industrial societies have become excessively complacent through surfeit and are now unable to muster the necessary strength.

Given the will, technology can serve in many crucial ways. Indeed, all the necessary advances in agricultural productivity, energy utilization, population control, conservation and recycling of materials, improvements in housing and transportation, presuppose technical input. No advances will occur in these areas until they are technologically accomplished.

The principal difficulties may lie in obtaining agreement to act. Pollution control will not happen unless there is concerted insistence by the majority of the people that it is necessary. Again, technologists probably will have to take the lead in convincing a majority of the necessity. To make advances through difficult social terrain, incentives as strong as those during wartime may be required. How can such incentives be instilled? Most people can be convinced either by the proofs of science or the authority of scientists that man is currently engaged in a life-and-death struggle with himself and the biosphere. To instill this conviction with sufficient intensity to assure support for further action, however, requires the strongest and most relentless sort of action. So far, few technologists have been willing to take the trouble to lead such a crusade for the attention of mankind.

It would be tragic if men awake to their peril too late. *The Sleep of Reason* was the title C. P. Snow chose for a book. It came from Goya's line for one of his Caprichos: "El Sueño de la razón produce monstruos." "The sleep of reason brings forth monsters." Historically, scientists and engineers have not recognized the duty of constantly serving as bugle boys to prevent the sleep of reason and the triumph of monsters. But such persistent reveille has become important.

Through their training, engineers learn that design decisions always involve a choice among alternatives. They learn that seldom if ever can they make an improvement in one direction without a loss in some other direction. Nature demands a price for each gain made. The design questions then become: What do we want most? What price can we pay? Technologists can transmit this lesson not only for technological choices but for general decisionmaking as well. Making the public realistically and broadly aware of the available alternatives, the liabilities and costs of each, could be the most fruitful role technologists play in society.

In July, 1976, at the age of 88, America's distinguished economist and science critic Stuart Chase wrote in *The New York Times:*

> One stands in profound awe of the human knowledge, skill and computation that sent a spacecraft half a billion miles for eleven months through space to a predetermined landing in a relatively benign area on the planet Mars. Then, like any tourist, the Viking took pictures and sent them back to friends.
>
> Is there any way we can apply this tremendous mental power to the achievement of some kind of steady-state society on earth, which our planet desperately needs if it is to survive the triple threat of pollution, plutonium, and uncontrolled population? (65)

That is the question.

REFERENCES

50. Editorial, *The New York Times*, July 25, 1976, p. 16E.

51. Du Noüy, Lecomte, *Human Destiny*, Longmans, Green and Co., New York, 1947.

52. *Ibid.*, p. 186.

53. Krutch, Joseph Wood, *The Modern Temper*, Harcourt, Brace & World, Inc., New York, 1956, pp. 44, 47.

54. *Ibid.*, pp. 53, 56.

55. Watson, James D., *The Double Helix*, Signet Book, The New American Library, New York, 1968.

56. Cousins, Norman, "Who Owns the Ozone?" *Saturday Review World*, October 5, 1974, p. 4.

57. Rocks, L. and Runyon, R. P., *The Energy Crisis*, Crown Publishers, New York, 1972, p. 20.

58. Rose, David J., "Energy Policy in the U.S.," *Scientific American*, Vol. 230, #1, January, 1974, pp. 20-29.

59. Maugh, Thomas H. II, "Natural Gas: United States Has It if the Price is Right," *Science*, Vol. 191, No. 4227, February 13, 1976, p. 549.

60. Hayes, Dennis, "Conservation as a Major Energy Source," *The New York Times*, March 21, 1976.

61. Carter, Luther J., "Energy Policy: Independence by 1985 May be Unreachable Without Btu Tax," *Science*, Vol. 191, No. 4227, February 13, 1976, p. 546.

62. Ray, Dixy Lee, Interview, *Science*, Vol. 189, No. 4197, July 11, 1975, p. 126.

63. Abelson, Philip H., *Energy for Tomorrow*, University of Washington Press, Seattle, Washington, 1975, p. 78.

64. *Chemical Week*, Vol. 118, No. 13, March 31, 1976, pp. 30-31.

65. Chase, Stuart, Letter, *The New York Times*, July 30, 1976, p. A20.

QUESTIONS FOR REFLECTION

1. Consider and debate the forecast that when the collapse comes, those in the developed countries such as the U.S. will suffer most. Is the collapse inevitable? What human values will help us through to the other side?

2. In connection with nuclear weapons, does safety depend on the numbers available? As world dictator, what would you do if anything with atomic stockpiles?

3. It is said that the U.S. has no energy program, but merely a policy of stumble from one crisis to another. What sort of U.S. energy program would you implement, given the power? Where would you obtain funds and where would you invest them to develop new sources and to conserve old ones?

4. Which energy sources appear best for the present...for the near future...for the distant future? What should we do that we are not doing in connection with energy? What should we not do that we are doing?

5. What altered living conditions will be necessary for effective energy conservation and pollution control? How can individuals be persuaded to accept such conditions and to comply?

6. Would it be feasible for a house or a city block to sort waste into four or five categories? If so, how could this be encouraged and enforced?

7. Is the scientist or engineer better equipped to run society than the lawyer? Than the philosopher? Than the artist? Than the economist? Should any*one* run society?

8. How can technologists assist the general public in recognizing and swallowing the dictates of reality?

9. Should population control be voluntary or must stringent methods be used to prevent population disasters?

10. What can be done to utilize the resources of the human mind for benign survival of the human race in place of selfish concentration on narrow immediate ends?

7. Critics of Technology

"Before he died in combat in the last war, Richard Hillary found the phrase that sums up this dilemma: 'We were fighting a lie in the name of a half-truth.' He thought he was expressing a very pessimistic idea. But one may even have to fight a lie in the name of a quarter-truth. This is our situation at present. However, the quarter-truth contained in Western society is called liberty. And liberty is the way, and the only way of perfectibility. Without liberty, heavy industry can be perfected, but not justice or truth."

Albert Camus
Resistance, Rebellion, and Death

THEODORE ROSZAK

In both the past and present, science and technology have not lacked critics. Kinder critics such as H. M. Tomlinson, British author, have simply regretted that technological progress does not seem to bring men any closer to self-realization and understanding.

> To-day we can fly around the world, and count the journey in days, not years (Tomlinson wrote in 1931); and, even so, it is just beginning to dawn on us, with this new power to our hand controlled by levers described as fool-proof, that we do not know our whereabouts, precisely, but must discover it. The globe must again be encompassed; our explorers must be guided by stars Drake never saw in his heavens. An industrious New World, though equipped with machinery that could supply an automobile and an aeroplane to every villa, cannot help humanity in its new voyage of discovery to learn its whereabouts and a right course; that essential truth cannot be found with an output of standardized machinery which is even prodigious in its magnitude. . . Curiosity, with the aid of material science, can do no more. We must go now to the discovery of a world where aeroplanes and submarines cannot take us. The harder task is ours (66).

A leader among contemporary critics of technology is Theodore Roszak (67). Older critics made some distinction between science and technology, considering technology the application of science and thus the direct agency for "mechanizing" the world. Those critics aimed their heaviest barbs at technology, with occasional accolades for the "purity" of science and its worthy search for knowledge. Roszak criticizes both technology and science. He objects to the basic methods of scientific thinking and what he calls "the myth of objective consciousness."

According to Roszak, "objective consciousness" (i.e., the scientific method) divides reality into two areas. In one, the investigator tries to obtain knowledge without personal involvement or commitment (the In-Here). The other area is the Out-There. In conducting experiments, the scientist must divorce himself from emotion, achieve complete objectivity, and treat the Out-There as totally nonintellectual and inert. To assure detachment, the scientist invents machines to make measurements. Thus mechanism replaces the mind. Roszak thinks/fears that substituting machines for men is the final goal of science.

To support his view that science leads to inhumanity, Roszak lists a number of investigations in the appendix of his book, *The Making of the Counter Culture.*

These investigations include experiments on humans and animals, a U.S. Defense Department study of the possible effects of a nuclear war, research into the air-raid shelter program, the possible breeding of human replicas, and computer simulation of abilities normally considered human. In these investigations and others, Roszak believes that scientific ideology overshadows other ways of thinking. He warns about a science controlled by experts inaccessible to ordinary citizens, and among whom specialization prevents intelligent communication. Technology, he argues, crushes man under a mass of gadgets and machines that methodically and inexorably dehumanize him.

Roszak's "Counter Culture" consists of the young who oppose the technological society and seek alternate living patterns. Roszak takes the mystical view that man should actively be a part of nature, keeping his faculties open and attentive so that when nature speaks he will hear. He endorses magic and spends much time discussing the tribal shaman, who assists the people in receiving communications from nature. He contends that if one hears nature properly (without the interferences and distractions of reasoning), one can develop his most desirable human attributes. The imagination, vision, and creativity become unchained, and man realizes his full potential. This is a familiar mystical concept. "If the doors of perception were cleansed," wrote poet William Blake, "everything will appear to man as it is, infinite." Roszak does not want completely individualistic reception of nature's signals. He feels instead that common experience should be shared through the tribal shaman.

The counter culture will not replace the present culture. Rather it will develop to blunt the worst features of the present culture and to explore new possibilities. In an interview, Roszak said that he had no wish to abolish science, merely to end its dominance. In his Utopia, presumably some sort of technology will provide for the needs of his counter culture. He assumes that people will have sufficient goods and the time needed to create the counter culture. Since rational planning has little place in his scheme, it is not surprising that Roszak pays little attention to man's temporal and physical requirements and ways of supplying them. Such requirements would be less perhaps when man achieves a more symbiotic relationship with nature.

In later publications, Roszak takes a mellowed and less condemnatory view of rationality (68). He admits the claim that rationality is part of the human spectrum. Roszak calls this spectrum the "Gnosis." Science has tended to corrupt the gnosis, he feels, in that like Frankenstein it has created a monster of meaninglessness. Though technologists mean well, they produce soulless work, which lacks the gift of love. Roszak believes that man wants to find the meaning of his existence and a revealing clue to his true identity. Without this meaning, work is not worth doing and science has no value. In a sense, this echoes the earlier, quieter critique of science provided by Tomlinson.

Roszak protests the reductive nature of scientific research which provides detailed information about nature, but not its essentials. He wants more synthesis, less analysis. Nature, he contends, holds the meaning of life. Man should seek that true meaning instead of building his own artificial meanings, through a science that increasingly views nature as indifferent and valueless until it is translated into pragmatic benefits.

In Roszak's spectrum of knowledge, there is a critical difference between information and meaning.

> At one end, we have the hard, bright lights of science; here we find information. In the center we have the sensuous hues of art; here we find the aesthetic shape of the world. At the far end, we have the dark, shadowy tones of religious experience . . . here we find meaning . . . Science is ... part of this spectrum. But gnosis is the whole spectrum.

Roszak considers that science has a role to perform if it can "join knowledge with wisdom," and that there is hope the work of technologists will serve the gnosis. If science does not accomplish this, man must discard it in the future.

Comments on Roszak's Thesis

A number of objections have been made concerning Roszak's view and indictment of science. First, it is doubtful that the science-engineering community has ever achieved anything approaching the dominant influence he attributes to it. Considering the administrators of large enterprises, whether business or government or church, the top positions are seldom held by scientists or engineers. These decision-makers are generally educated in the liberal arts areas, with perhaps professional training in law or business. Such leaders profit from technological advances, but they are seldom familiar with scientific methodology. Serious criticisms, in fact, have been made in this book and elsewhere that scientists and engineers have been insufficiently involved in political and other crucial decision-making areas. Far from using the power available to them, technologists rather have tended to remain amoral and unconcerned with the end results of their efforts. Roszak might with some accuracy accuse them of being dehumanized by their work; yet when scientists have become aroused, as they have in connection with nuclear energy, the SST, and pollution, their contributions have tended to be liberal and humanistic.

Clearly the scientific approach can be used for evil as well as good. Scientific inquiry is one avenue for dealing with human behavior. How the methods of science are used, whether for the benefit or the subjugation of man, depend on various human considerations, including politics. Because of criticism by Roszak and others, there is now wider recognition that amoral, disinterested science no longer suffices. Scientists and engineers must inform men concerning the facts of technology; they must persuade them to make the best decisions.

Roszak's central objection is to the methods of science and to the use of rationalism and objectivity. Scientists contribute to this objection themselves by repeatedly emphasizing that they deal with facts and not values. Those who make this claim delude themselves, since they inevitably make value judgments at every step. Truth, the fundamental credo of science, has the corollary that a false statement must be rejected. This ethical view enters into both the process and the result of the scientific effort. Scientists who pretend that facts alone are their concern have forgotten the assumptions upon which they work and need to be reminded. Critics such as Roszak deserve thanks for compelling such reassessments and self-reminders.

Creativity in science has a character similar to creativity as described by Roszak. Thus, arriving at a new principle may require 99 percent perspiration and 1 percent inspiration, as Thomas Edison suggested in a newspaper interview, but the 1 percent inspiration distinguishes the genius from the mechanic. In science, inspiration usually comes only after laborious thought and work. When the mind has been prepared, the seed of creativity has a chance to germinate and grow. It would seem to be little different with poetry, art, and music.

The poet Saint-John Perse was once asked by Albert Einstein how a poet works. Perse described how the idea of a poem came to him and grew, with an important part being played by the intuition and the subconscious. Einstein replied with delight, "But it's the same thing for the man of science. The mechanics of discovery are neither logical nor intellectual. It is a sudden illumination, almost a rapture. Later, to be sure, intelligence, analysis, and experiments confirm (or invalidate) the intuition. But initially there is a great forward leap of the imagination." (69).

Admitting as much does not submerge the foothills of science in the fogs of mysticism. Instead it recognizes a useful fact concerning the human mind, its astonishing capacity for such forward leaps, with no need to tremble when there are mystical overtones. The scientist does not wish to banish mysticism, as Roszak, like Martin Luther, so determinedly wants to banish reason. Luther wished to enthrone faith in the place of reason, but Roszak less convincingly prefers magic.

Roszak rightfully notes that humans contain intricate bundles of feelings with irrationality often prevalent. He considers it essential to avoid drowning these feelings in a sea of reason (though the reverse to some may seem the greater danger). The feelings of individuals range widely, of course, and it is difficult to see how all of these diverse impressions can provide a stable guide for society as a whole. Mystic experiences may be genuine for the individual, but sharing them with others may be impossible. If gifted shamans interpret the mystical signals from nature, all may be well for a time; but what will happen when one shaman's signals conflict with those of another shaman?

Can Roszak guarantee that the policies resulting from the insights of shamans will give people more freedom, insight, understanding, and the ability to create and grow? It seems doubtful that such a guarantee is possible. Indeed, historically, great decisions have often been made on the basis of hunches, guesses, fears, hates, visions, and mystical insights—all quite in harmony with the shaman approach—and often the decisions proved catastrophic. Today, irrationality competes tirelessly against rationality, with neither permanently gaining the upper hand; but considering recent American history from Vietnam to Watergate, irrationality, fear, hunch, hate, and perhaps even shaman magic often seem far more the architects of decisions than reason. Considering how detrimental and even disastrous, some of these decisions have been, suspicion becomes conviction that definite benefits would ensue from applying methods to test the correctness of beliefs, to have one person double-check another, and to test policies by the standards of reason as well as the "gut feeling" they impart to politicians and other decision-makers. Feelings were in command when they burned Massachusetts "witches" and lynched Blacks in the American South. Roszak's shaman might intend to bring men better into tune with nature, but the apprehension is never far away that sooner or later the shaman would see fit to instruct his followers in hating and smiting the shaman's enemies.

While saying this in defense of reason, there are still the nagging questions raised by Roszak. Does the technologist eventually tend to view and treat humans as objects? Does he lack concern for feelings and emotions because they do not match the neat and orderly criteria of facts? Roszak contends that we cannot understand the psychology of human beings through science. He advocates abandoning control of society by scientific means. Society obviously has never been scientifically controlled, but Roszak raises a valid concern about the extent to which such control can be managed. The technologist would suggest that much has been learned about the physical world by scientific means and that social problems could perhaps also be fruitfully approached in a like manner. If such an approach is taken, technologists obviously must have clearer guidelines to and sharper awareness of human needs than they bothered with in the past. Roszak's counter culture might supply valuable warning beacons along the way.

Ideally, what may be needed is a "united culture" approach rather than the divided, two-culture approach. C. P.Snow, who provided early warnings about our cultural divisions, considers it possible to work for and perhaps achieve such a union.

> With good fortune . . . we can educate a large proportion of our better minds so that they are not ignorant of imaginative experience, both in the arts and in science, nor ignorant either of the endowments of applied science, of the remediable suffering of most of their fellow humans and of the responsibilities which, once they are seen, cannot be denied (70).

This trial collaboration, leading to a permanent one, between technologists and humanists is needed without delay as a therapeutic reaction to the criticism of Roszak and his disciples. In human matters, "the road between" often proves easiest and most productive to travel—the road between reason and feeling, the road between scientific method and intuitive wisdom.

HERBERT MARCUSE

Herbert Marcuse was the leader in establishing the early philosophical stages for student antipathy to the technological society. This paved the way for later student unrest and radical action. Marcuse was educated in Europe and began his work there. Widely read and quoted, he was already becoming a sage and patron saint to youth when he came to America. His most influential book is probably *One Dimensional Man* (71) in which he seeks to reconcile the contradictory ideas of Marx and Freud.

Freud sought to interpret man's actions by developing theories of the mind. He and his interpreters offered elaborate concepts of the subconscious, the unconscious, the id, and the libido (sex). Instincts of life, death, power, and sex were made the cornerstones for modern psychoanalysis. Freud's theories have been widely used by critics of technology.

Freud stressed mind, while Marx and his followers tended to be skeptical of psychic exploration, which seemed to them intangible and bordering on the mystical. Herbert Marcuse's appeal to the young may derive in part from his attempt to combine aspects of both Marx and Freud (72).

In his critiques of technology, Marcuse holds that through technology, the advanced industrial civilization has subjugated men without most of them being aware that it has taken place. By supplying man's material wants and inventing other "false" needs, the technological society stifles any impulse toward radical protest, and it smothers all trends toward change (73). According to Marcuse, "the tangible source of exploitation disappears behind the facade of objective rationality." In *Eros and Civilization,* he wrote:

> Concentration camps, mass exterminations, world wars, and atom bombs are no "relapse into barbarism" but the unrepressed implementation of the achievements of modern science, technology, and domination. And the most effective subjugation and destruction of man by man takes place at the height of civilization, when the material and intellectual attainments of mankind seem to allow the creation of a truly free world (72).

Marcuse apparently does not consider technology a force in itself, but rather as essentially a means to domination. He does not directly ask the question who wields the means and who chiefly benefits from the use of technology for domination. Marxists simplistically define the beneficiary as the "ruling class" which reaps the rewards from the utilization of technology. Marcuse does not follow up these class ideas, being more interested in the impact of technology on mankind as a whole.

What remedies does Marcuse propose in connection with the domination by technology? As with many critics, he concentrates more on analysis than advice. Of course, it is easier to tear down a shaky structure than to determine ways of strengthening it. And ultimately, Marcuse is pessimistic about man's future. Had he witnessed our current energy crises, inflationary debacles, and other problems, he might have been encouraged about the potential demise of technology. Yet more likely he would have interpreted them as minor, temporary setbacks. Marcuse sensed an irreversible trend in which technology would be utilized inexorably for human domination. "The instruments of productivity and progress," he wrote, "organized into a totalitarian system, determine not only the actual but also the possible utilizations. At its most advanced stage, domination functions as administration, and in the overdeveloped areas of mass consumption, the administered life becomes the good life of the whole, in the defense of which the opposites are united. This is the pure form of domination" (74).

Hope he saw only in those who would resist the system, go into the streets without arms or protection and reject totalitarian answers. Those who did so would make themselves potential victims, but they alone, the revolutionaries against one-dimensional society, could reverse the direction of that society. This became the rallying cry for the students of the 1960s, but Marcuse warned them in advance that there was little expectation of success. "Nothing indicates that it will be a good end," he wrote, yet still his inclination was "to remain loyal to those who, without hope, have given and give their life to the Great Refusal" (75).

Comments on Marcuse's Thesis

Herbert Marcuse, the same as Theodore Roszak and other critics of technology, tends to endow the enemy with an omnipotence that reality fails to justify. Marcuse focuses with magnifying lenses on certain 1984 aspects of technology so that bugs become as large as monsters. It is natural enough to see one's monster as leviathan, fire-breathing, and invincible. But the philosopher who indulges in such exaggeration may despair unnecessarily. If a "one-dimensional man" is the product of modern society, technology cannot accept exclusive credit. Man himself is still substantially master of his fate, at least in those regions where a hostile environment does not totally command.

The limitations of technology are also given insufficient attention. Quite serious questions are now being asked about the future of the technological society because of material shortages, and it is increasingly obvious that man socially and politically will determine his technological future by the choices he makes now in connection with resources, pollution, and population.

Marcuse invited those who feared the one-dimensional, totalitarian fate to "drop out" or to make what he called the "Great Refusal." One generation of students made the effort. In the U.S., these students helped terminate the Vietnam War, they raised still unsettled questions concerning prevailing human values, and they eliminated certain hypocrisies from American society. Under their pressures, frankness and truth became more fashionable. Increased tolerance of different life-styles was a further by-product.

Technology, however, was little if any affected by the controversies that raged. It turned out that the "Great Refusal" did not include automobiles, television, efficient plumbing, computers, and all the other dispensations of modern technology. If technology had been one of the original targets, it soon proved futile to denounce it with the very methods and media of technology.

Marcuse was wise in his concern, and it would be injudicious to dismiss that concern without admitting that if technology were omnipotent, he would be correct in much of his argument. Technology has no omnipotence, however, and its limitations tend to keep it a miniature rather than a large-scale dragon. There is no disputing that some individuals and groups try using technology to bring about the totalitarian tyrannies Marcuse warned about. A parallel fact, however, is that technologists are even more interested in resisting such tyranny than mankind in general. There can be a one-dimensional man in the sense Marcuse hypothesized, but there cannot be a one-dimensional science. In their more important work, technologists long ago were forced to join the "Great Refusal" against unchallengeable dogmas, and had to accept the indispensable necessity of change. Technology can be a misused force, but in their commitment to the spirit of free inquiry, technologists at least are allies of critics such as Herbert Marcuse.

JACQUES ELLUL

Among the most voluminous and frequently quoted criticisms of technology is Jacques Ellul's *The Technological Society* (76). Our present society is characterized by "technique," according to Ellul, a member of the Faculty of Law at the University of Bordeaux, France. Ellul, using the word with overpowering repetition, insists that technique has taken over man's activities. Technique is defined as "the organized ensemble of all individual techniques which have been used to secure any end whatsoever." Individual techniques exist in such areas as economics, politics, work, propaganda, sport, and medicine. Ellul feels that man has become so concerned with technique, perfection of technique now overrides the ends desired, and technique itself becomes an end rather than a means.

The requirements of technique become so important, the state must organize all aspects of life with resulting decline of human rights and dignity. Technique dissociates man from the purpose of his work and alienates him from an intimate oneness with society. Technique enslaves science. Education becomes a technique for compelling man to conform, sports a technique to keep him appeased, and politics a technique for preventing revolution. Ellul makes no distinction between socialism and capitalism. Each uses the same techniques for its own purposes.

Why is technique an evil to be shunned rather than a blessing to be welcomed? Ellul feels that technique and humanism are incompatible. Technique is like Gresham's Law: The bad drives out the good.

In the process of learning about man (an outcome of technique), the less human man becomes. Ellul is pessimistic about stopping or slowing the advance of technique. Like a tidal wave, technique inexorably advances, with humanity helpless before it.

Ellul offers no solution. His idea of Utopia can be inferred, however, from what he dislikes. In this Utopia, society might be static as it was in the Middle Ages. The tensions and pressures of technology would be gone. Simplicity would prevail as in the Middle Ages when a man walked more, lived more slowly (though a much shorter life), and thus presumably was freer and had more time to contemplate (the man rushing to catch a train is not free, by Ellul's standards). In this world of reduced technique—technique is never entirely escaped since life itself is a technique of advancing from one day to another—man could become himself again, much as he was in the earlier times of religion and relative tranquillity, before modern technology.

Ellul's book is a reservoir of material for opponents of technology. His picture of society resembles that of Aldous Huxley's *Brave New World,* though with a more pedagogic flavor. Acceptance of his thesis would lead one to dropping out or another "Great Refusal" as with Marcuse.

Like many critics of technology, he glamorizes the past while decrying the present. He favors the uncomplicated periods of the so-called Middle Ages between the Fall of Rome and the Industrial Revolution, discounting the physical hardships, while emphasizing the spiritual serenity. Assuredly, during the reign of feudalism, few men were burdened with the consequences and responsibilities of intellectual freedom. If a man was freer to contemplate, *what* he would contemplate was rigorously prescribed by the Church and the Landlord. For every man with sufficient energy to manage spiritual reflection, there were numerous slaves or serfs who lived as near beasts of burden to support the privileged class. Perhaps such a system endowed a few members of the nobility and the clergy with favorable lives, but at the cost of grinding hardship for the many.

C. P. Snow pointed out that the "industrial revolution looked very different according to whether one saw it from above or below . . . To people like my grandfather, there was no question that the industrial revolution was less bad than what had gone before. The only question was, how to make it better" (77).

It seems probable that Ellul and othes who see immeasurable advantages in times past imagine themselves among the privileged rather than the exploited. Such enthusiasts for the past make the argument that serfs were contented with their lot, since they did not frequently rebel, and by enjoying the certainty of knowing their place in life, were happier than modern man who is constantly pulled from above and pushed from behind. The argument does not convince. The serfs under Nicholas II in Russia did not fight to save their Czar. Blacks in the American South did not refuse freedom when it was offered.

Ellul offers much intriguing criticism of technology, but ultimately his objections to "technique" have a semantic hollowness, as if they were full of sound and fury, but signifying little.

UTOPIAN DREAMERS AND CRITICS

Utopia or Technological Hades? The question has been asked about most of the fictional or philosophical Utopias that have been presented either seriously or satirically during the past century.

The concept of Utopia as a place where everything is perfect was not invented in this century. The word, derived from Greek roots meaning "not a place," was used by Sir Thomas More in his 1516 romance concerning an island where mankind's troubles cease thanks to perfect laws, politics, and morals.

Imagining a place where life is perfect is among man's oldest preoccupations, going back as far as views of Heaven in Biblical times, or Plato's Republic with its Philosopher King, to a wide range of modern Utopias from Samuel Butler's Erewhon to Orwell's Oceania in *1984* ("perfection" in Oceania would be Big Brother's view, not Orwell's!).

The dream of Utopia derives from man's dissatisfaction with his present condition and from the impulse to hope that present sufferings and sacrifices will somehow achieve a blissful future for mankind. Planning for the future is one of the characteristics distinguishing men from other animals, creatures of instinct imprisoned (or liberated?) permanently in the present world.

Edward Bellamy's highly successful 1888 novel, *Looking Backward: 2000 –1887,* idealized a technological Utopia. Bellamy describes society as he imagines it in the year 2000 in a Socialistic Boston. The author, a lawyer turned newspaperman, used the book as a method of communicating his social ideas that a planned technological economy could assure peace and security. He first intended the work to be a "cloud-palace for an ideal humanity," but the book awakened the social consciences of authors and many others. Influenced by his own work, Bellamy after 1888 became an active social reformer.

In his Utopian society, individuals achieve leadership positions by starting at menial but necessary tasks and progressing through good service to higher positions. The society has no classes, and everyone is equally supported. Nationalism, with everyone devoted to the service of the state, is one aspect of this Utopia, but Bellamy also stressed individualism achieved through religion.

Bellamy in 1888 was not critical of technology. In his Utopia, industrial progress serves to liberate mankind for humanitarian pursuits. This reflected the nineteenth century faith that through technology, all men eventually could enjoy well-being and richer lives (78).

In the twentieth century, however, as technology was turned to the modernization of war and to the growing regimentation of men, fictional Utopias switched from celebrating man's future to a satirical or despairing view of that future as suggested by the extension of technological trends.

The best known twentieth century works on this theme are Aldous Huxley's *Brave New World* (79) and George Orwell's *Nineteen Eighty-Four* (80). Each book contained an anti-technology bias, through an extension of the dark side of technology as it existed at the time the author wrote. Huxley's book was published in 1932, Orwell's in 1949. These highly popular novels have been much reprinted and widely distributed. Both continue to fan the flames of anti-technology, and it is a rare criticism of modern technology that does not quote extensively from one or both. The fact that Huxley wrote during a worldwide depression, with the decline of the west being predicted with melancholy emphasis by Oswald Spengler and others, is not always mentioned as background for understanding *Brave New World.* The fact that Orwell wrote after World

War II, in a shattered Europe, with Russian totalitarianism assuming command over half the continent, also helps explain the author's projections. In other words, political and social concerns quite as much as technology haunted each writer, though critics of technology have not conscientiously bothered to explain.

In *Brave New World,* Huxley describes a society seven centuries in the future. Controlled by experts with high mentalities and few human emotions, the general population works a regulated amount of time, has complete economic security, and is totally manipulated. Childhood conditioning, drugs, hypnosis, and media management are among the approaches designed to assure that each person is unquestioningly satiated with happiness and acceptance of his lot. Marcuse later echoed this idea by claiming that technology allows all protests to be absorbed.

In Huxley's society, an ovum is artificially fertilized outside the mother's womb and then methodically treated with nutrients and drugs to assure "scientifically" that a person of a specific type with controlled traits is "decanted" (born). Gene selection controls the numbers in various human classes from the Alphas (highest in intelligence and power) through the Beta, Gamma, Delta, and Epsilon classes. Society discourages individuality and suppresses free will while providing for all physical needs. War is eliminated. Reading and writing, except for propaganda purposes, do not exist. The upper classes indulge in sex games and orgies; the laboring classes are kept docile and contented with "soma," a drug producing the pleasures of intoxication without alcohol's aftereffects.

In a discussion between the Controller, a defender of the society, and Savage, a critic desiring some of the values omitted from the society in the interest of stability, the Controller insists "it would upset the whole social order if men started doing things on their own." "Ford forbid," he declared, that man should get the idea of "doing things." This would undermine stability, and "instability means the end of civilization."

> The greatest care is taken to prevent you from loving any one too much. There's no such thing as a divided allegiance; you're so conditioned that you can't help doing what you ought to do. And what you ought to do is on the whole so pleasant, so many of the natural impulses are allowed free play, that there really aren't any temptations to resist. And if ever, by some unlucky chance, anything unpleasant should happen, why, there's always *soma* to give you a holiday from the facts ... Christianity without tears—that's what *soma* is.

When Savage, representing the age-old impulse of humanity against the mindless ease of Utopian regimentation, says that he wants to experience everything ostracized from the society—danger, freedom, goodness, sin—the Controller observes that he is in fact "claiming the right to be unhappy."

As Huxley pessimistically saw the future in 1931, science would be so triumphantly and antiseptically in command, any effort to experience the emotional fullness of life would end in failure, as does Savage's effort. In 1931, Huxley did believe that the social consequences he foresaw resulting from the triumph of technology were far in the future. Twenty-seven years later he feared that *Brave New World* was taking shape much sooner than he had expected.

> The prophecies made in 1931 are coming true much sooner than I thought they would. The blessed interval between too little order and the nightmare of too much has not begun and shows no sign of beginning. In the West, it is true, individual men and women still enjoy a large measure of freedom. But even in those countries that have a tradition of democratic government, this freedom and even the desire for this freedom seem to be on the wane. In the rest of the world freedom for individuals has already gone, or is manifestly about to go. The nightmare of total organization, which I had situated in the seventh century After Ford, has emerged from the safe, remote future and is now awaiting us, just around the next corner (81).

Huxley took the name of his book from Shakespeare's *The Tempest,* "O brave new world,/ That has such people in 't." Some aspects of his Utopia he also borrowed from a speech of Gonzalo in that same play, written more than three centuries before Huxley's book, and anything that could reasonably be called modern technology. Gonzalo in his "perfect commonwealth" would allow no letters, "riches, poverty, and use of service, none." "No occupation; all men idle, all;/ And women too, but innocent and pure; No sovereignty,—"

Shakespeare in turn had obviously taken the attributes of his Utopia from an Essay by Montaigne, written and published in the sixteenth century. Montaigne in his scheme for a Utopian Republic set the pattern for Gonzalo by omitting sources of social friction such as knowledge of letters and the presence of magistrates.

The point is that designers of Utopias, from the early ones through Huxley, tended to make them sybaritic, self-indulgent places; Heavens with no useful work to do ("What a hell of a place heaven would be with no work to while away the time," we can almost hear someone like Montaigne or Mark Twain exclaiming.).

Satirically blaming science for such developments was Huxley's inspiration, but the suspicion rises that he found many aspects of his Utopia in the sixteenth century rather than the future. In other writings, Huxley more specifically and accurately indicted technology:

> We see, then, that modern technology has led to the concentration of economic and political power, and to the development of a society controlled (ruthlessly in the totalitarian states, politely and inconspicuously in the democracies) by Big Business and Big Government. But societies are composed of individuals and are good only insofar as they help individuals to realize their potentialities and to lead a happy and creative life (82).

Huxley gave poor grades not only to the society of *Brave New World* as he imagined it, but to the societies of the 1950s as he saw them. Finally, though, on reflection, we wonder if technology as Huxley criticized it was not more the symptom than the cause. There is an impulse to paraphrase Cassius from another play by Shakespeare: "The fault, dear Brutus, is not in our technology,/ But in ourselves, that we are underlings."

The author of by far the most influential and frightening of modern "Utopian" novels noted with searing clarity that the fault lies in us, in our technology, and in our remorseless fate. George Orwell in *Nineteen Eighty-Four* portrays a dismal anti-Utopia in which the population is totally controlled by the Thought Police and their leader, Big Brother. The society is continually at war with a remote enemy (the U.S.-Russian cold war was beginning in maddening earnest when Orwell wrote). The people do not know the reasons for the war. They do not need to know. During Hate Week, the Ministry of Truth efficiently leads them in what and how to hate. And daily all join in the Two Minutes Hate exercise.

Orwell's book was a despairing prediction that totalitarianism would inevitably be victorious and would submerge humanity, probably forever. Erich Fromm, the philosopher-psychiatrist, wrote of *1984:*

> George Orwell's *1984* is the expression of a mood, and it is a warning. The mood it expresses is that of near despair about the future of man, and the warning is that unless the course of history changes, men all over the world will lose their human qualities, will become soulless automatons, and will not even be aware of it (83).

Dr. Fromm elsewhere, expressing his own views concerning the effects of technology on individuals, wrote:

> Our contemporary Western society, in spite of its material, intellectual and political progress, is increasingly less conducive to mental health, and tends to undermine the inner security, happiness, reason and the capacity for love in the individual; it tends to turn him into an automaton who pays for his human failure with increasing mental sickness, and with despair hidden under a frantic drive for work and so-called pleasure (84).

In Orwell's 1984 society, some of the things concerning Fromm were swallowed up in doublethink, Big Brother's protective, eternal watchfulness, and the Ministry of Love where all social aberrations could be expertly "cured." Thus, Winston, the rebellious hero with deviant wishes for love, thought, and individualism, is treated at the Ministry of Love, and emerges to realize with total certainty that he loves Big Brother. That was the terror Orwell saw at the end of the road man was travelling, everyone loving Big Brother without question, without dissent, even lacking the words of dissent or complaint or dissatisfaction thanks to "newspeak," the language of 1984. Paralleling absolute love for Big Brother was automatic hatred on cue, every day without fail. "The horrible thing about the Two Minutes Hate was not that one was obliged to act a part, but that it was impossible to avoid joining in."

Orwell did not commit the error of accusing technology alone for Big Brother. He noted only that modern technology made Big Brother's success easier, and made it possible for him to achieve and preserve total power. Individuals were helpless against authoritarianism with technological control.

Orwell's book had perhaps greater political and social impact than any other Utopian or anti-Utopian book ever written. The language of *1984* entered general speech quickly. The meanings of "doublethink," "newspeak," and the ubiquitous caption, "Big Brother is Watching You," are still quite clear even to nonreaders of the book. The book hit an exposed nerve in modern society, and the pain has not left us. Not yet, because 1984 is too close, both chronologically and metaphorically. Huxley's effort at literary prophecy in *Brave New World* is read as literature and by those interested in curious psychological and biological ideas, but *1984* is still the book watching us "asleep or awake, working or eating, indoors or out of doors, in the bath or in bed—no escape."

"The threat of 1984 has roosted like a vulture in a tree, most firmly in the American consciousness," wrote Lawrence Malkin in a 1970 essay, "Halfway to 1984" (85).

Orwell did not live to see the development and playing out of the cold war between East and West, but it is unlikely that he would have been surprised at developments in the West, paralleling and seeking to balance similar or imagined developments in the East. Vance Packard in *The Naked Society* commented on modern electronic surveillance (he wrote before 1975-1976 revelations concerning U.S. CIA and FBI intelligence operations with a Big Brother ominousness about them) and concluded that "If Mr. Orwell were writing his book today rather than in the 1940s his details would surely be more horrifying" (86). Technological advances, in short, have increased the scope for monstrous possibilities in 1984.

The antithesis may also be true, that technological advances have served to stiffen resistance to the possibilities of 1984. Orwell, writing in a time of Stalinism, surrounded by symptoms of Western totalitarianism beginning to develop, was not optimistic. He quoted Nietzsche's warning and believed it: "He who fights too long against dragons becomes a dragon himself: and if thou gaze too long into the abyss, the abyss will gaze into thee." Yet Orwell all his adult life was a crusading journalist. He did not issue his warnings after it was too late. He wrote *1984* with no great expectation that his warnings would be heeded, but nevertheless, the hope lingered that humanity rather than Big Brother would win.

Orwell might be encouraged by the student uprisings against the establishment in the 1960s. He certainly would be encouraged that those who abused electronic technology by spying on others in the 1960s and 1970s were exposed, ridiculed, and maybe, at least for a time, stopped.

Perhaps the reality of technological strides in electronics and other areas have made individual citizens somewhat warier than before. Perhaps. Time will tell, but it does seem now, at least in the U.S., that "1984" is a little farther away than 1984. It seems that though Big Brother may have been watching, he too now is being suspiciously watched in turn. It even seems that though we gazed long at dragons, we turned back just short of becoming one and gave ourselves another chance at humanity. In 1965, British statesman, Gladwyn Jebb, lecturing at Columbia, said of Orwell, ". . . it must surely be admitted that from the purely international point of view we have already come quite a long way toward realizing his predictions . . . the system Orwell contemplated for 1984 is already to a large extent in existence" (87).

That was 1965. The cold war went on. The Vietnam confrontation was a continuing crisis. Rapprochement with China had not begun. The Twenty-Four Hour Hate was prevalent on all sides. And *detente* was a word only international relations professors had heard. At that time 1984 seemed close indeed, but it wasn't technology that had brought it close.

A decade later, much has changed. With technology advancing at a dizzying pace—men on the moon, Viking on Mars—still 1984 totalitarian horrors have receded. Could it be that technology is not responsible for *1984* and *Brave New World* societies when and if they happen? Could it be that human values determine human societies more surely than machines?

A MISCELLANY OF CRITICS

Charles Reich earned success in 1970 with his book, *The Greening of America* (88). Reich foresees a paradise in the U.S., achieved with little planning or effort. How? Like Roszak, he advocates a counter culture of men and women who pay attention to the instincts, obey the rhythms of nature, and who set aside that troublesome tyrant, the rational intellect. Reich urges each of us to begin with the "self," to accept this as the "only true reality," and to turn inward. Reich's faith in the greening process relies on mysticism rather than technology. He tills in much the same soil with the same tools as Roszak, and acknowledges his debt to Marcuse and Ellul. In the student movement of the 1960s, he saw America breaking loose from its old commitment to the corporate state, and a new consciousness appearing "like flowers pushing up through the concrete pavement . . . For one who thought the world was irretrievably encased in metal and plastic and sterile stone, it seems a veritable greening of America."

Reich, though critical of the uses to which technology has been and can be put by the corporate state, does not criticize technology *per se.* He sees it instead as one of man's liberating influences potentially:

> The crucial point is that technology has made possible that "change in human nature" which has been sought so long but could not come into existence while scarcity stood in the way...That which we called "human nature" was the work of necessity—the necessity of scarcity and the market system. The new human nature—love and respect—also obeys the law of necessity. It is necessary because only together can we reap the fruits of the technological age. And it is necessary because only love and human solidarity can give us the strength of consciousness to withstand the overwhelming seductions and demands of the machine (89).

A long-time outspoken critic of the American educational system is Robert M. Hutchins, former President of the University of Chicago. As President of the prestigious Center for the Study of Democratic Institutions, Hutchins aimed barbed thrusts at American education's preoccupation with technical specialization. When a university president, he was quoted to the effect that college graduates receive sheepskins to clothe their intellectual nakedness, and in 1945 he stated that "We do not know what education could do for us, because we have never tried it." Education to Hutchins must include familiarity with mankind's classical literary heritage. An engineer or physician, whose specialized training prevents this familiarity, may become a competent technician, but he is not educated. Hutchins insists that absolute concentration on a speciality to the exclusion of all else crams the student with facts and more facts, but gives him no time to correlate, synthesize, or judge (90). Because of the monopolistic rigors of specialization, Hutchins sees a contradiction between the reality and the factors stressed by scientists as vital aspects of scientific method, such as open-mindedness, intellectual honesty, a will to know the truth, curiosity, objectivity. He doubts that scientific procedures can help us decide what we want to be and the kind of world we want. To escape our present specialization treadmill, he advocates major changes in educational methods as well as many more adult study centers and opportunities (91).

Hutchins influence has been substantial. His ideas have helped bring about more "cultural" courses in engineering curricula. The results as yet have not been impressive, perhaps because of a casual approach on the part of some faculties, and because most engineering students have not been helped to see how such courses relate to their professional needs. The importance of continuing and hopefully improving such courses is not challenged by the adults, as opposed to the "calculators" on university campuses. In time, when engineers recognize that human values are the most important component in design, they will gradually become involved in broader issues than the specialized ones of their discipline. Active debate on alternatives in design classes may stimulate this involvement. The time for such debates will be hard to come by, however, since the stuffing of facts is such an entrenched custom. One further fact might, if sufficiently highlighted, prove convincing: The fact that computers can store facts quite competently, freeing the human mind for creativity. Perhaps eventually engineering and scientific students will be "educated" before they are graduated. When it happens, thanks will be due to Robert Hutchins and other gadfly critics of technical specialization.

Kingsley Davis, Professor of Sociology, University of California at Berkeley, argues that social science cannot be included in the same category as the physical sciences due to the difficulty of control in the social field. Most solutions to social problems are simple to postulate, says Davis, but sometimes impossible to apply. In the case of cigarette smoking, for instance, it is established that cigarettes are injurious to health, and warnings have been widely circulated. Still smoking increases. Those who smoke do so knowing the warnings. Seat belts in automobiles are another instance of obvious safety protection that millions knowingly ignore.

Technology can find solutions to problems (e.g., seat belts in automobiles), but it has a much poorer record persuading people to apply a solution, especially if it involves any inconvenience or change of familiar habit.

Davis contends that the absence of solutions cannot often be attributed to a lack of knowledge, but rather to ineffective application of knowledge. He details what social science can accomplish in providing essential information and interpretation concerning problems and their solutions (92). More is involved than simply telling people "how." They must also be shown and convinced "why." In his effort to make the social sciences more pragmatic in their response to human needs, Davis does not criticize technology. He simply indicates that in his view technology cannot, by itself, solve basic human problems.

Writers have long found technology exceedingly available as well as convenient to criticize and blame in whole or part when they observe something wrong. From Stuart Chase's *Rich Land, Poor Land* (93) to Stewart Udall's *The Quiet Crisis* (94), technology has taken the proverbial rap along with human neglect for the exhaustion of resources, deterioration of the environment, pollution, and the pressures of population. Paul Ehrlich stressed the continuing dangers of runaway population growth in *The Population Bomb* (95). Technology has even been assigned responsibility for too many babies! (It keeps them alive!)

Ecology and environment have entered the working vocabularies of average citizens. Problems involving these concerns persuade Alan Wagar in a *Science* article, "Growth versus the Quality of Life" (96), that growth itself must henceforth be carefully questioned. Growth, as an unquestionable good, has been an almost sacred principle in America, with realtors and businessmen leading the worship services at its shrines. But it is increasingly recognized, despite great reluctance by many and great resistance by some, that many factors limit growth, including available land and resources. No automobile on earth can move on oil that has already been consumed.

Listening to the critics, it seems manifest that future growth for mankind must lie in quality rather than quantity. If the quality of life is to be enhanced, technology will be essential and most critics know it. They stress the sins committed by technology, always through the selfishness, caprice, ambition, or ignorance of men. This is not, of course, condemnation of technology. C. P. Snow, Charles Reich, Stuart Chase, Aldous Huxley, and many other estimable observers of man and his works have not hesitated to admit that if man's lot on earth is to be improved, the contributions of technology are indispensable.

Yet even those who praise the potential contributions of technology to man's welfare and freedom, emphasize the problems of using technology constructively, and they often criticize technologists for not helping more in seeing to the benevolent application of their work. It is no doubt healthy for scientists and engineers to be aware that they are not universally adored.

REFERENCES

66. Tomlinson, H. M., *Out of Soundings,* William Heinemann Ltd., London, 1931, pp. 70-73.

67. Wade, Nicholas, "Theodore Roszak: Visionary Critic of Science," *Science,* Vol. 178, December 1, 1972, pp. 960-962.

Roszak, Theodore, *The Making of the Counter Culture,* Doubleday and Co., Inc., New York, 1969.

68. Roszak, Theodore, "The Monster and the Titan: Science, Knowledge, and Gnosis," *Daedalus,* Summer, 1974, pp. 17-32.

69. Fuller, R. Buckminster, Walker, Eric A., and Killian, Jr., James R., *Approaching the Benign Environment,* pp. 134-135.

70. Snow, C. P., *The Two Cultures,* p. 91.

71. Marcuse, Herbert, *One-Dimensional Man,* Beacon Press, Boston, 1964.

72. Marcuse, Herbert, *Eros and Civilization,* Vintage Books, New York, 1962.

73. Marcuse, Herbert, *One-Dimensional Man,* p. 32.

74. *Ibid.*, p. 255.

75. *Ibid.*, p. 257.

76. Ellul, Jacques, *The Technological Society,* Alfred A. Knopf, New York, 1964.

77. Snow, C. P., *The Two Cultures,* p. 31.

78. Bellamy, Edward, *Looking Backward: 2000-1887,* Houghton Mifflin, Boston, 1926.

79. Huxley, Aldous, *Brave New World,* Harper & Row, New York, 1948.

80. Orwell, George, *Nineteen Eighty-Four,* Harcourt, Brace and Company, New York, 1949.

81. Huxley, Aldous, *Brave New World Revisited*, Perennial Library, Harper & Row, New York, 1965, p. 4.

82. *Ibid.*, p. 20.

83. Fromm, Erich, "Afterword," *1984*, Signet Classic, The New American Library, 1961, p. 257.

84. Fromm, Erich, quoted in *Brave New World Revisited*, p. 20.

85. Malkin, Lawrence, "Halfway to 1984," *Horizon*, XII, No. 2, Spring 1970, p. 37.

86. Packard, Vance, *The Naked Society*, Pocket Books, Inc., New York, 1965, p. 25.

87. Gladwyn Jebb, Lord Gladwyn, *Halfway to 1984*, Columbia University Press, New York, 1966, pp. 4-5.

88. Reich, Charles A., *The Greening of America*, Bantam Books, New York, 1971.

89. *Ibid.*, p. 415.

90. Hutchins, Robert M., "Stamp Out Engineering Schools," *Professional Engineer*, March/April, 1968.

91. Hutchins, Robert M., *Science, Scientists, and Politics*, Center for the Study of Democratic Institutions, Santa Barbara, California, 1963.

92. "Research in the Service of Man: Biomedical Knowledge, Development, and Use," Committee on Government Operations, U.S. Senate, 90th Congress, 1st Session, 1967.

93. Chase, Stuart, *Rich Land, Poor Land*, McGraw-Hill Book Co., New York, 1936.

94. Udall, Stewart L., *The Quiet Crisis*, Holt, Rinehart & Winston, New York, 1963.

95. Ehrlich, Paul R., *The Population Bomb*, Ballantine Books, Inc., New York, 1968.

96. Wagar, J. Alan, "Growth versus the Quality of Life," *Science*, Vol. 168, June 5, 1970, pp. 1179-1184.

QUESTIONS FOR REFLECTION

1. Is man rational or irrational? Which is better? Why?

2. In consideration of the critics, should educational programs for scientists and engineers be revised, and if so, how?

3. What should the technologist do about his critics: Go on about his business? Listen and learn from what they say? Wait for them one at a time on a dark night? What, if anything, can the technologist learn from critics and what should he do about it?

4. In what way is it fair to blame technology for modern ills?

5. In what way is it not fair to blame technology?

6. How can Shakespeare, Montaigne, and *Alice in Wonderland* help a technologist do his job, if they can?

7. What *is* the technologist's job as a citizen in society?

8. Advocates of Technology

"One machine can do the work of fifty ordinary men. No machine can do the work of one extraordinary man."

"Keep away from that wheelbarrow! What the hell do you know about complicated machinery?"

Elbert Hubbard

FAITH OF THE DEFENDERS

Technology is not adrift on a log in a sea of critics. There are supporters, and not only those who benefit from or live by technology. Thinkers of stature who have observed the development of the modern age since the seventeenth century include many who seek to praise technology, not to bury it.

Faith in technology has been vigorously tested in the twentieth century, more so than in any earlier period. Two hundred years ago, critics of the slowly developing industry in Europe and the British Colonies tended to be more concerned about offending the Creator of the universe than the visible dangers of industrialization. Some pious and superstitious folk worried that it was dangerous to tamper with the ordained reality of nature. It was "ordained" because "that was how it was," and the status quo could be altered only with grave risks. Thus, when Benjamin Franklin announced details of the lightning rod he had invented, readers of his 1753 *Poor Richard's Almanack* in which instructions for building such a rod were given, were predictably divided in their reaction. Some acclaimed the invention as progress. Others shuddered and warned that lightning was God's instrument of wrath and should not be thwarted by such diabolical instruments as protective rods. Fear of lightning rods, then steam engines, steamboats, railroads, telegraphs, and telephones, gradually became jettisonable in the eighteenth and nineteenth centuries when no signals of raging disapproval were received from higher sources.

As fear departed, faith in a sense replaced it. The belief developed that the works of technology in the main were for the good of man and to be welcomed with enthusiasm. The twentieth century, however, has confronted that faith with new realities that seem bent on restoring fear. Two world wars and recurrent "minor" wars have been powered and equipped by technology, with efficiency so total that more than one nation can now technologically destroy mankind.

In most of the currently industrialized nations, the fruits of technology grow from a poisoned tree . . . pollution, environmental degradation, the earth stripped of its mineral wealth, human beings crowded into suffocating and demeaning cities. At least, these are ways that critics of technology have seen fit to describe the modern world and to charge scientific progress with the general blight.

As a result, defenders of technology have found themselves on the defensive. They have been forced to answer critics instead of methodically building a case for technology that would convince the judge (history) and the jury (mankind) of the defendant's innocence, good will, and value to society. In a recent *Science* editorial, Gerald Edelman of Rockefeller University wrote:

> Basic science is now on the defensive. It is being assailed by groups and governments as being costly and dangerous, as being silly or ominous. Now that modern biology, for example, has achieved some mastery over genetics, various scenarios of deliberate intrusion into the human gene pool are being rehearsed, almost always with more anxiety than insight. These fears reflect our society's failure to understand the fundamental process of basic research and its relation to our historical and legal heritage. As a result of this failure, the ever-growing influence of scientific invention and technology has terrified us as much as it has given us peace (97).

Science has been taken into the family, though to some members it seems a mad eccentric with unpredictable pills, potions, and explosives in every pocket. Science must be tolerated, since it keeps the house going after a fashion, but it is rather like living with a mysterious stranger who may at any moment run amok.

A number of careful observers have tried to explain the truth about the mysterious stranger and to eliminate the fears that cause terror. Their testimony for the defense is worth considering.

J. BRONOWSKI

Bronowski was a distinguished biologist intimately concerned with human values and their relationship to science. His book, *Science and Human Values* (98), revealed its author as a philosopher-scientist with the gifts of a poet. Almost single-handedly Bronowski did much to bridge the culture gap reported by C. P. Snow. At least he formed convincing beachheads on both shores.

In the preface to *Science and Human Values,* Bronowski indicated that his purpose would be to demonstrate that a practitioner of science must form a set of universal values. His book was inspired by a visit to Nagasaki, Japan, one of the two cities on earth so far with the unique distinction of having been substantially demolished by an atomic bomb.

The book consists of three lectures, entitled: "The Creative Mind," "The Habit of Truth," and "The Sense of Human Dignity."

Bronowski does not hold either the scientist or the artist responsible for our current predicament. Yet he writes that neither in justice can absolve himself from blame.

In the first lecture, Bronowski reemphasizes that the acts of creativity are essentially the same whether accomplished by the scientist or the artist. He refutes the misapprehension that scientists are little more than meticulous collectors of facts. "Science is a search for unity in hidden likenesses," he writes, and stresses that *interpretation* of facts is the critical part of science, not the gathering. The classic events of science illustrate this point. Newton did not finish his work by reporting to the Royal Society of London and scientists around the world that an apple had fallen from a tree. That was the fact, and to the average observer, that would be all: An apple fell. But the scientist could start from the simple fact, and pursue an elaborate chain of reasoning and inspiration to the universal laws of gravity and motion. Was Newton merely collecting facts? Or using them with genius?

Newton interpreted stray facts to reach mammoth generalizations that serve scientists daily as established principles: All bodies in the universe, from the largest to the smallest, are attracted to one another, by a force that can be measured and expressed as a gravitation constant. Newton's principles offered a firm foundation that served more than two centuries for a generally accepted model of the universe and theories of cosmogony. Newton had not discovered all truth—what scientist does—and later science determined that Newton's laws were not quite accurate in the case of electron motion. But he brilliantly found some truth.

Another fact or two needed interpreting. Albert Einstein appeared to do the job. Sir Richard Glazebrook wrote that Newton's "mechanics guided astronomers and men of science in their search for natural knowledge. And if in these last years Einstein has carried us some steps further, has picked up some few more of the jewels of truth, which Newton sought on the shore, Newton's laws remain, included it may be, in a more comprehensive statement of the truth."

Newton and Einstein interpreted the facts available to them and provided models of the universe to fit known criteria. No doubt there will be further models and new ways of looking at the universe as additional facts extend the statement of the truth. Do scientists merely collect facts? Reflect on Isaac Newton and the fact of a falling apple.

Bronowski compared the tools of creative thought used by writers and scientists. He finds that similar conceptual images are used. Both artists and scientists are concerned with understanding nature, and in their work must reach far beyond a mere collecting of data. Creative imagination is the essential common tool used by artists and scientists.

In his second lecture, Bronowski examines the differences in the creative acts of art and science. Science proceeds normally in three steps: 1) collection of data, 2) determining the order in the data, and 3) establishing the unifying concept that explains the data. Thus, Johannes Kepler used the planetary observations of others as well as his own to discern order and regular orbits in planetary motion. Steps one and two were accomplished. Then Isaac Newton completed step three with the unifying concept: the universal law of gravity.

Following steps 1, 2 and 3, a scientific discovery or hypothesis must be confirmed in all available ways. If Newton's model of the universe was correct, certain behaviors should result which could be checked. Newton did not know the cause of gravity, or why two masses attract each other. But his concept proved extremely useful in practical applications. Later Einstein proposed fresh concepts of mass and gravity and constructed a new model using abstract mathematics. And again, confirming tests had to be performed. Did Einstein's model accommodate known facts? Essentially it did, though modern physics is continually trying to learn more of the truth, which might set off a new wave of model making. And so on, ad infinitum, as knowledge adds to knowledge, and as science rigorously tests every hypothesis.

As the chain of knowledge proceeds from bare facts, to principles, to conceptual models, credence is given to Bronowski's insistence that "the symbol and the metaphor are as necessary to science as to poetry," and that science is a creative rather than a mechanical process.

The difference between scientific concepts and concepts involving social relations, including ideals, lies in the classical view that nonscientific concepts cannot be tested empirically. The idea persists that such concepts as justice, honor, and dignity are values derived from some absolute and ideal source. Religions have always supported these views. Yet Bronowski points out that although our prejudices follow such patterns and are relatively immune to the lessons of experience, our value concepts have not remained fixed, but have changed as knowledge of the world has changed. Bronowski shows how concepts of self-interest were replaced by concepts of the classless society, and how such value concepts are altered by experience. Similarly, notes Bronowski, the theories of science change when more subtle tests show previously missed flaws. Bronowski concludes by maintaining that the world of what is and the world of what ought to be must in both cases be subject to experience and truth, just as our value concepts are subject to experience and truth.

In his third lecture, Bronowski discusses the human values used by science. He insists that truth has the same meaning in the arts as in science, though art sometimes has greater difficulty in communicating its truth. He refutes the logical positivist stand on values, though logical positivism has been called the method of science. Positivists have argued that whatever *is* may be subjected to tests, but what ought to be cannot be tested. The fallacy of logical positivism lies in the assumption that only known truth can be tested and that tests can be conducted independently of society.

Scientists, Bronowski makes clear, must rely on others to reach conclusions, they must make and apply continual social assessments—as a consequence, value decisions enter at every step in science.

The scientist (as Newton and Einstein admitted) depends first on the work of those who preceded him. Scientific knowledge has grown slowly through the ages, and has built up communally. Trusting that knowledge (the work of others) is essential. The scientist depends as well on instrument makers, his associates, his peers. All are important to him in choosing a direction for his work and making progress in the direction chosen. He is inevitably affected by his dependencies when he makes value judgments concerning which experiments to start, which data to keep, and which to discard.

In a letter from Monticello on June 18, 1799, Thomas Jefferson, champion of science who made a name for himself in another field, wrote: "Science can never be retrograde; what is once acquired of real knowledge can never be lost . . . as long as we may think as we will, and speak as we think, the condition of man will proceed in improvement" (99).

Stressing the unity of science across time as well as geography, J. Robert Oppenheimer envisioned a "house of science" where all scientists can "go in and out; even the most assiduous of us is not bound to this vast structure. One thing we find throughout the house: there are no locks; there are no shut doors; wherever we go there are the signs and usually the words of welcome. It is an open house, open to all comers" (100).

To assure the effective use of the house of science, a reliable system of values is needed. Bronowski demonstrates how science has constructed the values it needs. Every human effort is fallible, and errors must be corrected. Part of science's value system must stress the obligation to correct errors and ways to do it. Human dignity must be preserved among the values of science, and Bronowski writes: "Those who think that science is ethically neutral confuse the findings of science, which are, with the activity of science, which is not." He contends that the chief evil of wrongdoing lies not in the harm the wrong itself may cause, but in the wrong that others may be encouraged to do in imitation, thus loosening the cement of society.

Bronowski illustrates how the international society of scientists has developed ethical codes and managed to survive political upheavals. He believes, as do most scientists, that freedom and justice have become abiding goals of mankind partially through the efforts—and needs—of science. Science must have freedom to seek the truth, and then to express it, if only in the minds of scientists within the confiding walls of the house of science. Fiction may be written about science, but ultimately there is no fiction in science, only truth and the freedom to find it. Thus other men might force a scientist to recant a truth to save himself, as men of the Church did Galileo; but legend would have it that the old scientist walking away was heard to mumble, "E pur si muove." But it does move. Whatever he recanted, whatever others said, truth was truth, and his mind was free to know it.

Even in *1984*, Big Brother could triumph. There might be no escape. But there was still, with great effort, one place left to every man: "Nothing was your own except the few cubic centimeters inside your skull" (101). A free scientist would recognize the value and vastness of that territory.

Bronowski implies the difficulty of retaining ownership of the human mind when he contends that a serious modern dilemma comes from the inability of human values to control mechanical science. Nevertheless he finds science more humane than governments. The tolerance of science does not guide nations, and Bronowski notes the intensity of self-interest in governments. He believes that science has outgrown such concerns (102).

Bronowski's third lecture in this valuable book ends on the note that science is a human activity and thus should not be distinguished from other human activities in which imagination plays a central role. The painter or poet expresses values as profound as those of science. The artist's methods of exploration may be different in some ways, but the parameters of exploration are the same and require the active presence of comparable human values.

In "The Ascent of Man" television series, Bronowski showed how science and art are interwoven in man's development. He used the cathedrals of Western Europe to illustrate the creative union of art, architecture, and engineering. He showed how optics had assisted astronomy, and how both optics and astronomy had allowed the development of celestial navigation and liberated seamen to explore every corner of the globe with no apprehensions about finding their way home again. Throughout the series, we saw scientists as humans with fears and faults, rather than detached beings absorbed in the cold facts of inhuman laboratories. Under Bronowski's guidance, it was reassuring to remember that great science was accomplished in earlier times, before specialization, by men whose curiosity made them scientists but who earned their livings in many other ways. The scientific ranks included Philadelphia printers such as Benjamin Franklin, clergymen such as Joseph Priestley and Ebenezer Kinnersley, and even a Superintendent of the British Mint (Isaac Newton). The extensive list offers a lesson to the modern world of technology in the fact that amateur scientists for the most part laid the foundations of today's science.

Bronowski devoted time to making his audience aware that science undergoes constant self-analysis. A scientist never quite accepts an answer as final. A theory is never quite absolutely established beyond the reach of possible alteration. New data may require altering any theory, or abandoning a cherished notion entirely. The chronology of science is littered with facts that turned out not to be accurate and with theories that proved wanting at some fresh turn in the road. Workers in technology expect this. They know the inescapable necessity of constant re-analysis. Bronowski makes clear that misguided, absolutist, religiously dogmatic principles—not science—lead to insanities such as the destruction of Jews by Fascism.

Bronowski's title, "The Ascent of Man," informs us that he believed man was progressing toward greater achievements in both technology and his own human aspirations. Bronowski encouraged a reawakening of hope for man's technological and human future. His optimism contrasts markedly with the pessimism shared by the critics of technology, and he insisted on a direction opposite to that of Roszak who seemed convinced that only artists are gifted with imagination and that rationalization must be abandoned before man can free himself from technology and reach fulfillment.

Bronowski also contrasts with those who consider values absolute and inherent. He suggests that values develop as men need them, that scientific endeavor itself became a recognized human value through the need for the benefits of technology. Those benefits have been essential for mankind to grow and prosper both materially and culturally.

Concerning the methods of science, Bronowski thinks they could usefully be exported to other disciplines. He sees them contributing to orderly and progressive relations in politics, society, and other fields.

Doubts can be expressed concerning some of his ideas. For example, in testing the results of an action in the social sphere, what standards do we employ? Are there standards?

Nevertheless, faults aside, it is regrettable that Bronowski's ideas have not received more widespread attention. The reason Bronowski failed to achieve this attention for his ideas may, ironically, be found in the fact that he argued in favor of technology as a force capable of "building up." He did not engage in the easier task of tearing down. The critic of society still receives more attention than the champion of society. A Newton hasn't appeared to explain why, but it happens. Bronowski offers considerable guidance for technologists as well as nontechnologists. For depressed moderns, he is essential reading.

MELVIN KRANZBERG

If Roszak is technology's dragon, Kranzberg is its Galahad. Since the 1960s, he has wielded a sharp sword in defense of technology, which he considers a vital contributor to the arts as well as humanity.

Unfortunately, Kranzberg, on the faculty at Georgia Institute of Technology, Atlanta (he was formerly at Case Western Reserve University, Cleveland), makes his points in lectures and addresses at meetings, and may not reach the large audiences found by Roszak, Marcuse, or Ellul (103). He may also suffer because sensational condemnation attracts more attention than reasoned support. Conjuring up devils to blame for bleak situations is easier, and more popular, than denying that there are devils, and purposefully suggesting less dramatic cures.

Kranzberg remains unperturbed, if he even knows that he may have a somewhat smaller following than some of his adversaries. Instead, he concentrates on attacking the anti-technologists and defending technology as a positive force. He points out initially that man himself is a result of technological change. The capacity for thought of *homo sapiens* appeared after the initial trial and error efforts of *homo faber* (man the maker). Man is first distinguished from other animals by his use of tools. The dialectical interaction of toolmaking and thought has continued to the present day. It is the union of toolmaking and thought, of the practical and the theoretical, says Kranzberg, that provides man with the skill to move mountains, and the inclination.

He shows how the need for irrigation in Italy lead first to the siphon pump, which was unexplained until Torricelli advanced the theory that air has weight. This in turn lead to Boyle's law, and from this countless practical developments came. Simultaneously in Italy, speculations concerning planetary motions started a chain of events leading to Newton's laws.

Throughout man's history, the humanistic impulse has distinguished the human from the brutal. For much of that history, technology has been part of man's development, and inevitably has shared in man's humanistic goals. Technology as well as humanism is exclusively human.

High among man's humanistic goals has been that of relieving material want so that individuals would have greater time to develop their uniquely human attributes. Kranzberg believes individual freedom to pursue peculiarly human goals arrived at flood stage when the Industrial Revolution altered the productive mechanisms of society and thus changed the conditions of life. Machines first liberated man's muscles from long and exhausting toil. Automation extended this liberation as technology moved to more advanced stages. Today the computer has the capacity to free man further from repetitious tasks of a mechanical nature. Only as he has become free of such dull, necessary routines, has man reached the point where he can strive to be genuinely free in the political and social sense. The man who must spend many hours of his day performing rote actions that a machine or computer could do is not yet free, but rather is still a prisoner of past necessities.

Gerard Piel has said, "Slavery became immoral when it became technically obsolete." The machine and the computer have made it indisputably obsolete in a technical sense. In this sense, Piel views technology as a liberating force, both preparing mankind for freedom, and making freedom worth having.

Social democracy is clearly a by-product of technology, according to Kranzberg. Today the average person enjoys material comforts, art, music, recreation, and other benefits previously reserved for the wealthy. Today there are more concerts to attend, and more people attending them than ever before. Thousands of new books appear annually. World famous actors on television present theatrical masterpieces for audiences of millions, dwarfing the total audiences they could reach during a long professional career in theaters. It was said that a particular production of Shakespeare's *Merchant of Venice* with Sir Laurence Olivier reached more people with one performance than *all* the performances of the same play in theaters since Elizabethan times. No longer are opera and ballet the exclusive pleasures of a few. Millions experience them on television, on film, and in theaters around the world. Technologists, by designing and producing new conveniences and achieving beneficial breakthroughs, have done more to change social relations and liberate individuals than all the writers in man's history. Perhaps Shakespeare's contemporary, Francis Bacon, had intimations of this development when he wrote more than three centuries ago, "Books must follow sciences, and not sciences books."

In the course of fulfilling human needs, technology, somewhat oddly, gave man the leisure and the time to question and criticize his own creations, including technology. Originally identified with humanism, technology now is the focus of attacks by those working in the humanities. The popular warning that "there is perhaps no surer way of incurring a man's wrath than by doing him a good turn" may be too broad for universal application, but it does seem a relevant description of many humanists as the critical beneficiaries of technology.

Those in the humanities need a target for the disappointments and frustrations of modern times. Technology happens to be conveniently available. Technologists also need a target, for the same reason, and some of them have simply imitated those in the humanities and attacked themselves. Kranzberg will have none of this, including the fact that often now the word "humanistic" is used to mean "spiritual," with no recognition of the "material." Historical evidence demonstrates that mankind's distinguishing achievements are material creations, whether statues, sports palaces, books, or microtransistors. By equating humanism with the "spiritual" exclusively, man's creations are dismissed in favor of primal sensations—man should sink down into his own psyche and wallow in the multicolored muds. Kranzberg, among others, considers this nonsense and says so loudly. He argues that modern poets have let us down by failing to fulfill their traditional role of interpreting nature and society to man. His solution? Scientists and engineers and psychologists—all those who recognize that material accomplishments, bifocals for instance, are not evil per se—may have to fulfill the tasks themselves that poets were unwilling or unable to do.

Technology in itself, of course, is neither good nor bad. It enables, it does not compel. The most convincing critics of television, for instance, are not those who insist it is bad and thus should be destroyed, but those who simply say: Turn it off. Those who prefer the Technicolor shores of their own psyches, in the case of technology, can simply turn it off. They are not compelled to drive automobiles, or store their soybeans and cranberry juice in a refrigerator. They are not even obligated to use a ballpoint pen. Technology has set them free to do a wide range of things. And ironically, many of the most ardent critics of technology are simultaneously even more ardent beneficiaries of it.

Wide use has been made of Mary Shelley's story of Frankenstein to illustrate that man creates the mechanism of his own annihilation. Doctor Frankenstein built the monster to do good and thought of himself as a benefactor of man. Opponents of technology admit this, but claim that in spite of man's intentions, the creations of technology turn against man inevitably due to some flaw in technology. Your ballpoint pen eventually will have no choice but to squirt ink in your eye.

The central fact in the Frankenstein allegory (usage has made it that) is that the "monster" sought warmth, understanding, and sympathy at first and wished to reciprocate these qualities. But the monster was an outsider, he was different, he wasn't like everybody else—and his overtures of friendship were repulsed. He turned against man as an enemy only after being continually rebuffed. The moral of the allegory condemns in men whatever makes them intolerant of diversity, nonconformity, or any departure from the familiar mediocre and the triumphant average. Such intolerance is found, and not infrequently, even in countries with democratic pretensions. Essayist E. B. White wrote, "The concern of a democracy is that no honest man shall feel uncomfortable. I don't care who he is or how nutty he is."

Whether benefactor or monster, technology is clearly an indispensable reality. Mystics cannot will it away. Critics cannot drive it away. In truth, even the most sincere of the critics probably do not really want to exorcise it, but simply to control and direct, and in short, to "humanize" it. Technologists themselves are with accelerated frequency recognizing the need to add something more than technical facts to their education. If they do not realize it as undergraduates, they later realize their need for familiarity and understanding of humanistic matters. At that time many of them either launch themselves on a self-imposed regimen of reading and study, or they register for additional courses at the nearest liberal arts college. What about the "liberal artists?" Are they recognizing or admitting the importance in a technological world of understanding science and knowing about technology? Actually there has been a greater tendency for technologists to journey to the other side of the campus, than vice versa. Why this is true needs explaining, and efforts need to be directed at making the highway between the two cultures a two way street.

One of the things that makes Kranzberg as refreshing as a shower on a hot day is that he shouts his message with sufficient exuberance to be heard on both sides of the campus. He presents his case forcefully from a background of intellectual and historical research, but without the stridency of some economically or politically motivated supporters of technology whose ardent know-nothingism (or know-very little ism) does less good than harm to the cause of technology.

Kranzberg, like Bronowski, inspires hope, just as Roszak and Ellul on the other side force reflection. A debate between Roszak and Ellul on one hand with Kranzberg and Bronowski on the other would have been very interesting.

HARRISON BROWN

In one of his books, *The Next Hundred Years,* Harrison Brown, a Professor of Geochemistry at the California Institute of Technology, concludes that by using known techniques and resources, man can provide a reasonably abundant life for an increasing population. This optimism presupposes that the chief difficulty, which does not lie in technology or resource availability, but in the will of man to face his problems and do what he must, can be overcome (104).

One critical aspect of this difficulty is whether or not mankind will have the strength and intelligence to avoid war. Brown shows how difficult it would be to recover from such a conflict simply from a physical standpoint, without considering the devastating cost psychologically and other ways. He advocates remedial measures such as more drastic population control, greater reliance on international friendship, and cooperation with instead of resistance to the United Nations.

He feels that these challenges and the problems they bring can be met with intelligence. He admits that the major unanswered question is whether men will recognize the predicament and take the necessary steps. He hopes that new developments in psychology, such as human behavior research, will provide more people with understanding and the ability to control hostile feelings and destructive actions. This hope, however, extends deeply into the future, not the immediate present. Now is the time for necessary changes to cope with the most serious crises the human race has been required to face.

Brown chooses to be optimistic, partially because the situation is not, in his view, hopeless and because it is no great task to indicate the specific things that must be done. He is also optimistic because no useful reason exists to be otherwise. Either man listens to sanity and acts accordingly, or else.

R. BUCKMINSTER FULLER

Buckminster Fuller, inventor of the geodesic dome, the dymaxion house, and other concepts, can safely be described as an incorrigible optimist in his view of the future, who is certain technology will continue to serve the needs of human progress.

Fuller seeks to reform the environment instead of men. Man he believes should control nature, rather than the other way around. In his professional credo and actions, he echos lines written by the poet Matthew Arnold:

> "In harmony with Nature?" Restless fool,
> Who with such heat dost preach what were to thee,
> When true, the last impossibility—
> To be like Nature strong, like Nature cool!
> Know, man hath all which Nature hath, but more,
> And in that *more* lies all his hopes of good.
> ...Man must begin, know this, where Nature ends;
> Nature and man can never be fast friends.
> Fool, if thou canst not pass her, rest her slave.

Buckminster Fuller has never seemed inclined to rest, much less to rest a slave. The task he sets himself, and all technologists, is much too great for the luxury of despair, pessimism, or very much other than work.

> My task as inventor is to employ the earth's resources and energy income in such a way as to support all humanity while also enabling all people to enjoy the whole earth, all its historical artifacts and its beautiful places without one man interfering with the other, and without any man enjoying life around earth at the cost of another. Always the cost must be prepaid by design-science competence in modifying the environment (105).

The tetrahedronal city illustrates both Fuller's optimism and imagination. This city would consist of a pyramid structure, measuring about two miles in length along each side. It would be anchored in the ocean and would float permanently on the water. The tetrahedronal city would provide living accommodations for about one million people. It would generate power with atomic reactors and use the heat from the reactors to desalinate water supplies. This would be Fuller's method of coping with population growth—many such cities floating on the world's oceans.

Fuller's geodesic sphere is another bold concept. The sphere would be half a mile in diameter, and the air within would be heated by the sun, causing the structure to float in the air. Fuller envisions people living in such spheres. They would serve splendidly, he thinks, for transportation of men and materials. In man's future world, Fuller sees many such spheres, floating cities, domed cities, dwellings underground, and many other imaginative adaptations of earth's available space, both under and above the surface. Such innovations would leave the land free for agriculture and other useful purposes.

Fuller's distinguishing characteristic is his refusal—perhaps his inability—to be gloomy about mankind, or to curb his imagination concerning the future of man. He refuses even to be dismayed by the enormous challenge of population growth.

> ...despite the hullabaloo about the world population explosion—all of humanity could be brought indoors in the buildings of greater New York City, each with as much floor room as at a cocktail party. All the cities of our planet cover sum-totally less than 1% of the earth's surface (106).

The thing to do is to use fully what we have, creatively, boldly, imaginatively. Fuller sometimes views man as a unique, highly specialized machine (107) programmed and predesigned for success on earth. At the same time he sees man as endowed with a dynamic capacity for change. Fuller emits enthusiasm and stimulation as he supplies a constant flow of fresh ideas, but he seems little concerned about overpopulation, war, pollution, and hunger, contending that intelligent technology can cope with all of them. His Utopian schemes, unfortunately, though grandiose and intriguing, seem too farfetched to assist with immediate problems. Fuller might rightfully call this a paralysis of man's will or a weakness of his purpose. Paralysis of the will and human weakness, however, are realities that have to be confronted and resolved as we work our way toward the future. When the future is reached, perhaps Buckminster Fuller and his tetrahedronal cities will be waiting for us. In what spirit should we greet them?

OTHER ADVOCATES

If a plebiscite were held on the question whether technology should continue, those in favor would dominate if only from the votes of those who depend on technology for livelihoods, and those who lack technology on a significant scale but are certain they want it. Whether technology would do as well among intellectuals on university campuses and elsewhere is another matter entirely. Technologists might find themselves in a minority among their colleagues who don't mind accepting the benefits of technology, but avoid public commendations. User-critics of technology have often shown themselves contradictorily but cheerfully inclined to accept the comforts and conveniences of technology while protesting their intellectual purity.

There have been outspoken defenders of technology, and not all of them have worked for General Motors, Mobil, or Lockheed.

Bronowski, Kranzberg, and Fuller qualify as intellectuals by modern standards. Others have spoken with authority from a base of knowledge. Robert Theobald is one who has done his homework, and he stresses that all people have a claim to resources even though they may not contribute economically (108). Theobald argues that a sufficient material foundation must be achieved to eliminate want for all humans. He also is optimistic that it can be done, given the will to make the necessary human effort.

Robert L. Heilbroner, Professor of Economics at the New School for Social Research, wants to extend our technological responsibility not just to include everyone now living, but everyone who will live. His concern embraces our neglected and increasingly ignored posterity. If we knew with certainty that mankind would not survive a thousand years, speculates Heilbroner, unless we gave up our incredibly inefficient diet of meat, abandoned pleasure driving of the family automobile and in effect the family automobile, and reduced our use of energy to the minimum essentials, would we do it? "Would we care enough for posterity to pay the price of its survival?" asks Heilbroner. Observing present trends and assessing current attitudes, his answer is this: "I doubt it." But he does not believe current attitudes necessarily have to prevail. He asks the question: "Suppose that, as a result of using up all the world's resources, human life did come to an end. So what?" And he answers it:

> I am hopeful that in the end a survivalist ethic will come to the fore—not from the reading of a few books or the passing twinge of a pious lecture, but from an experience that will bring home to us...the personal responsibility that defies all the homicidal promptings of reasonable calculations. Moreover, I believe that the coming generations, in the encounters with famine, war and the threatened life-carrying capacity of the globe, may be given just such an experience. It is a glimpse into the void of a universe without man. I must rest my ultimate faith on the discovery by these future generations, as the ax of the executioner passes into their hands, of the transcendent importance of posterity for them (109).

Gerard Piel, publisher of *Scientific American,* in his *Science in the Cause of Man,* details some of the victories of science for the sake of mankind's past, present, and future generations (110).

Other optimists about the future of mankind, served by technology, include Glen Seaborg, Marshall McLuhan, and William Cozart. Prominent industrialist David Sarnoff, Board Chairman of RCA, earned the right to be considered one of the advocates when he analyzed the extent to which communications and computers can beneficially revolutionize man's control of nature and his own abilities. Sarnoff believed man should readily assign to machines any tasks they can accomplish, and thus reserve his own energies and talents for fuller self-development and achievement.

Some pro-technology writers—too many perhaps—in the vein of Buckminster Fuller have viewed the future in terms of more machines and elaborate technological gadgets. All these developments may take place.

Indeed a great many more than we can currently imagine are likely to appear. Still entirely accurate and appropriate for mankind's future are the words written by Benjamin Franklin in 1780 to chemist Joseph Priestley. Franklin was commenting on the great strides being achieved by science: "The rapid progress *true* science now makes, occasions my regretting sometimes that I was born so soon. It is impossible to imagine the height to which may be carried, in a thousand years, the power of man over matter."

Yet the future requires more than simply cleverer and cleverer machines. No impieties are committed in disputing the sacredness of machine technology and denying it permission to dwarf wider human concerns involving the greater questions of man's future (such as war, peace, population growth, and hunger). Technology will have vital contributions to make in answering the great questions and solving the difficult problems they bring. Those contributions are likely to be made with greatest effectiveness if we do not paint ourselves into a corner where only one subject dominates: rustproof and ineffably impressive gadgetry.

Give us the gadgets by all means. They will be needed to help in the higher tasks. But for the present, to enhance the survival chances of ourselves and posterity, we need more than visions of machines, or the counter vision of those who study technology and see only devils. More useful at the moment are the insights of Bronowski, Kranzberg, and others who can praise what merits praise without deifying. Also useful are the contributions of the environmental critics who have forced us to take a hard look at the mechanistic, stainless steel future, and to ask questions.

In compassionately and intelligently ordering technology for the future, both critics and advocates serve all of us well if they persuade us to ask questions, to look before leaping, to think.

REFERENCES

97. Edelman, Gerald, M., "Scientific Quests and Government Principles," *Science*, Vol. 192, No. 4235, April 9, 1976, p. 99.

98. Bronowski, J., *Science and Human Values*, Harper Torchbook, Harper & Row, New York, 1965.

99. Jefferson, Thomas, quoted in J. Robert Oppenheimer, *Science and the Common Understanding*, Simon and Schuster, New York, 1966, p. 110.

100. Oppenheimer, J. Robert, *Science and the Common Understanding*, p. 85.

101. Orwell, George, *1984*, Signet Edition, p. 26.

102. Bronowski, J., *Science and Human Values*, p. 70.

103. Selected Writings of Melvin Kranzberg:

"Technology and Human Values," *The Virginia Quarterly Review*, Vol. 40, Number 4, Autumn 1964.

"Man and Megamachine," *The Virginia Quarterly Review*, Vol. 43, Number 4, Autumn 1974.

"The Unity of Science-Technology," *American Scientist*, Vol. 55, #1, March 1967.

"The Disunity of Science-Technology," *American Scientist*, Vol. 56, #1, Spring 1968.

"The Technological Revolution and Social Reform," Chapter 3, 34th Yearbook of the National Council for the Social Studies.

"Confrontation: Technology," *Allegheny College Bulletin*, Winter, 1968-69, pp. 10-15.

"Engineering Education for Our Times," 52nd Commencement Address, Newark College of Engineering, June 6, 1968.

104. Brown, Harrison, *The Next Hundred Years*, Viking Press, New York, 1957.

105. Fuller, R. Buckminster, *Utopia or Oblivion*, Bantam Books, New York, 1972, p. 348.

106. *Ibid.*, p. 217.

107. Fuller, R. Buckminster, *Nine Chains to the Moon*, J. B. Lippincott, New York, 1938.

108. Theobald, Robert, *An Alternative Future for America*, Swallow Press Inc., Chicago, Illinois, 1968.

109. Heilbroner, Robert L., "What has posterity ever done for me?" *The New York Times Magazine*, January 19, 1975, pp. 14-15.

110. Piel, Gerard, *Science in the Cause of Man*, Alfred A. Knopf, New York, 1961.

QUESTIONS FOR REFLECTION

1. In what way is technology a good thing for modern man?

2. In what way is technology a bad thing for modern man?

3. In connection with technology, what can be done to buildup the good, eliminate the bad?

4. *How* should college students in arts and letters become more familiar with science and technology? *Why* should they become more familiar?

5. Which is the greater danger: Arts and letters people knowing little about technology, or technologists knowing little about arts and letters?

6. What ways would you suggest for both sides of the campus to cross the chasm between them and to commence an effective dialogue?

7. If you had to make a talk on the past benefits of technology, what specifically would you say?

8. If you had to discuss the future of technology, what would you say? Would your remarks be favorable or unfavorable?

9. What can be done to help political and social groups in your area appreciate, understand, and act on technical problems instead of clinging to ignorance and hoping against hope that the technical problems will simply go away?

10. If technology is considered a threat to mankind among some of your acquaintances, what can you do to inform them and yourself?

11. Should some scientific research be suspended in view of possible consequences, or should research proceed uninhibited?

9. Rationality vs. Irrationality

"*The true value of a human being* is determined primarily by the measure and the sense in which he has attained liberation from the self."

Albert Einstein, 1934

"What a lucky thing the wheel was invented before the automobile; otherwise, can you imagine that awful screeching?"

Samuel Hoffenstein

DETERMINING WHICH IS WHICH

How can you tell an advocate from an opponent of technology without reading his published works, and perhaps interviewing his mother about how he was as a child? Did he busy himself making new designs with Tinker Toys or did he chase butterflies with the somehow mystical idea of joining in the flight? The answers to these questions won't settle the matter, but they may offer a bit of evidence, which when added to other evidence, add up to a mystery inside a conundrum.

Physical appearance provides few useful clues. These days, the advocate as well as the opponent may wear glasses, or not wear them, have hair on his face, or not have it. Particular caution is required in distinguishing one from the other. What does he think? The key question—but be careful. Some opponents of technology think they are advocates, and some advocates think they are opponents. To start distinguishing, ask a few questions, listen to the answers, and perhaps certain giveaway signs will indicate which way a person is leaning.

There are a number of attitudes that opponents of technology tend to share. The absence of these attitudes lends fuel to the suspicion that one has an advocate of technology on his hands. The critics of technology, modern descendants of those in an earlier time who thought we were going to Hell in a Model-T Ford, tend to share a fondness for the past. They are especially sentimental about the good old days when life was simple and happy, before we messed ourselves up. Opponents of technology tend toward the mystical. They make a big to-do about listening to the instincts more attentively than to the intellect. They often stress the primacy of feelings and emotions over reasoning and rationalization. They choose intuitive solutions to problems, rather than meticulous analysis with detailed planning. Planning would involve excessive calculation. It might confuse the instincts, which will solve man's problems if not circumvented and betrayed by technical and rational complexities.

The modern opponent of technology often stresses accommodation with nature rather than control. Many who hold this view however, actively take advantage of technology's sovereignty over nature.

Advocates of technology praise the scientific method, though they may not be entirely certain what it involves. They esteem the use of reason, sometimes without using it. They believe that technology has helped more men than ever before achieve pleasanter lives than ever before, and their faith is constant that technology in the future will do even better. The advocates are optimistic about the future as technology with increasing efficiency controls nature for the benefit of man. Predictably, advocates are inclined to accentuate the positives, while opponents bemoan the negatives. The former point out how technology has reduced hardship in the world, eliminated slavery, substituted machine power for muscle power, and promises to free man to an even greater degree in times to come. The critics point to the frantic depletion of earth's resources, the spreading poison of pollution, the tension and sickness and misery of life for millions of people throughout the world, the loss of man's roots with the past and the earth, as his life becomes increasingly dehumanized and regimented.

There are abundant and convincing arguments on both sides. Technology obviously has benefited the human race in many ways and continues to do so. Also quite obviously the problems man now faces are attributable in large part to technology. The advocates argue that the benefits greatly outweigh the drawbacks and that the problems can be solved under man's direction. Opponents fear that man's ability to give such directions has been eroded by technological dependency, and that he lacks the will and the commitment to human values necessary to resist the easy but destructive temptations of technology. Advocates stress that technological progress cannot rightfully be faulted for man's own weakness, and that the challenges to existence from the natural environment and from within man himself cannot be ignored if the human race is to survive. This very question of survival represents a central part of the concern expressed by critics. With mounting threats to the biosphere, largely technological in origin, whether or not technological man will maintain a viable environment is a matter of great concern. Critics of technology point out that certain grim conditions are reached, and then defenders of the status quo argue that nothing should be done.

Thus, in the eighteenth century, for a time the owners of British ships carrying slaves from Africa to the West Indies argued successfully that nothing should be done to stop the trade. Why? Because the livelihoods of so many were involved. Today the automobile industry and others use the same argument—that the whole economy would be jeopardized if pollution requirements are made too stringent. So nothing should be done.

Opponents of technology bemoan the tradition that allows industrialists to make a virtue of profits whatever the environmental or human costs. Advocates insist again that this is a human, not a technological question. They note that technology's utilization is a product of human decisions as directed by human values. So the problem is man himself.

Back and forth the arguments roam, like a tennis ball at Wimbledon, and it often seems that the debaters are considering entirely different subjects, with the judges unable or unwilling to publicize the fact. One side talks about spiritual matters, the need for man to rejoin his roots, the obligation to stop poisoning himself and the earth. The other side talks about liberating mankind from ancient fetters and limitations. Oddly enough, both sides talk about freedom, but on opposite sides of a mountain.

Is the mountain too high to climb? It seems necessary to try crossing from one side to the other. Goal: To help the debaters begin considering the same questions with compatible definitions and greater coherency.

Is technology good *per se,* or bad *per se?* Did man leave something behind in the past that he should struggle to recover? Is the spiritual side of man being eroded by the acids of technology? Is rationalism a frightful disease to be struggled against. Does irrationality give natural man a better chance to survive with grace?

These are some of the persistent questions considered in pursuit of the truth about technology, and simultaneously the truth about man.

RETREAT TO THE PRESENT

When the present is demanding, tense, and uncertain, the future looms much the same way. Far from offering a sanctuary, the future is ominous with hard challenge and seems to contain the difficulties of the present magnified many times. That explains the impulse during periods of stress to gaze longingly toward the good old days of the past. It is easy to romanticize the past. The past is complete. Whatever was bad about it need not be repeated. And the good can be unblushingly exaggerated to the heart's content.

Was the past better than the present?

Despite the most sentimental efforts of the impulse to nostalgia, if confronted by the facts of history, an explorer in time has difficulty making a case for the superiority of earlier ages over the present. If the past held less urban tension, more cohesive and peaceful family relationships, and greater serenity for some, there was disease, economic anxiety, and dismal hardship for the many. Nostalgia seeks refuge naively in the good old days, which were notoriously characterized by the constant prospect of calamity from disease or the elements. In *Leviathan,* Thomas Hobbes writing during the seventeenth century described the life of man as "poor, nasty, brutish, and short."

The protections, both medical and physical available in our technological civilizations, were missing then. Every individual was vulnerable in numerous ways largely alien to the citizens of the modern industrial states. Today's citizen may have other problems—such as the threat of human extinction— but it cannot be said that Hobbes' description still fits the beneficiary of technogical advances.

The threatening past, of course, can be found closer to our own time than the seventeenth century or earlier. We do not need to rent a time machine and journey to the Middle Ages for confrontation with the plague, smallpox, or diphtheria that killed thousands and kept the life expectancy low by contemporary standards. Even those under 30 in many American cities such as Pittsburgh can remember when coal-burning furnaces created hazardous pollution with virtually no one except an occasional irritated housewife raising a note of complaint. The furnaces represented "prosperity" and "progress," and went unchallenged until recent years. Few would care to repeat such wanton conditions of pollution or other atrocious health conditions that characterize much of our history from the recent past backwards. We glamorize Colonial days in America, yet among those emigrating to America from England in the early period, less than 25 percent survived the voyage. Among the celebrated Pilgrims who settled at Plymouth in Massachusetts, more than half died during the first winter, and corn was planted on their leveled graves the following spring to deceive the Indians.

A modern version of retreating to the past has been the effort of some harried moderns to drop out of the general society and escape to the wilderness. How does this sort of venture fare? It turns out that most cannot exist without current-day technological support on a large scale. G. Ray Bane, Alaskan writer on wildlife, advises those on the "back to nature kick" that the rustic cabin allowing a return to the natural state doesn't exist (111). Bane points out that very few really have the stomach for living in a truly primitive manner, and his realistic appraisal makes clear that the "noble savage" envisioned by Rousseau and others serves a romantic illusion rather than a practical reality. Bane says that wherever man introduces himself, the wilderness is altered and displaced. One after another, comforts are brought in, and the simple life recedes. When Bane visits isolated people he finds that their first concern is to learn what's going on in the outside world. Their minds and emotions rarely have "dropped out" at all.

Romanticizing the past and the simplicities of wilderness living illustrates the natural impulse to avoid confronting present difficult issues. Yet obviously the problems of the present cannot be solved in the past, and the retreat to either the past or its corollary, the wilderness, merely postpones the inevitable grappling with whatever nightmares of reality or spears of rancor await mankind. A continuing task for those with judgment and perspective is persuading others to open their eyes before it is too late, to stir from their traditional apathy before the threats to human security become too powerful for successful combat; in short, to promote a "retreat to the present," where the problems for man are located and where they must be solved.

MYSTICISM AND RATIONALISM

In a popular Broadway musical by Rodgers and Hammerstein, the King of Siam sings a song with this philosophical lament: "In my head are many facts of which I wish I was more certain. Is a puzzlement." "Puzzlement" seems an apt summation of man's experience in connection with the spiritual side of his nature, and despite all the efforts of science, the puzzlement continues. There have been periods in modern history when some observers decided that religions slowly were being replaced by the sciences. As human technology advanced and as science probed somewhat deeper into the unknown, the various faiths gave the appearance of slowly retreating in both the numbers of their adherents and the scope of their philosophical territories. It was a philosopher who wrote, "Where there is the necessary skill to move mountains there is no need for the faith that moves mountains" (112).

Yet faith has persisted, and the religions even with diminished territory and power have prevailed. In some areas perhaps, as in the popular spread of interest in oriental religions, or in the so-called "Jesus Movement" of the 1970s across America, Europe, and other parts of the world, there have been signs of a religious renaissance. The ancient questions about man's spiritual nature are still puzzlements.

Bertrand Russell contended that religion has on balance produced more bad than good (113), but he also admitted that what science has to offer "has not satisfied the philosophers" (114). Religions still deal with the "puzzlements" in various ways that are considered valuable by many, including some scientists. Physicist J. Robert Oppenheimer, for instance, did not shy away from the word faith when he wrote of "our binding, quiet faith" in knowledge, and there are religious echoes in his hope that science could assist in bringing "a little light to the vast unending darkness of man's life and world" (115).

The effort to find a crossroads where science and religion can meet still goes on. Some scientists such as Eddington and Teilhard de Chardin, the Jesuit priest, have considered it an inevitable intersection at the point where science can travel no farther and the journey of religion begins. Other scientists have been less certain of the crossroads' location but quite certain that it exists. The remarkable Einstein in theoretical physics went farther than any man before him, and his mind was not enclosed within limited horizons. To Einstein the scientific mind and a religious feeling were profoundly related, yet not in the naive, orthodox fashion of most institutionalized faiths. "The scientist is possessed by the sense of universal causation," wrote Einstein, ". . . His religious feeling takes the form of a rapturous amazement at the harmony of natural law, which reveals an intelligence of such superiority that, compared with it, all the systematic thinking and acting of human beings is an utterly insignificant reflection. This feeling is the guiding principle of his life and work . . .It is beyond question closely akin to that which has possessed the religious geniuses of all ages" (116).

Eddington thought that scientists are drawn in their work by an inspirational light "as truly as the mystics" (117). Many scientists, committed to reason and rationalism as essential tools for solving man's problems, tend to be impatient with what they skeptically consider fuzzy and not particularly useful thinking in observations such as this by Eddington: "It would be wrong to condemn alleged knowledge of the unseen world because it is unable to follow the lines of deduction laid down by science as appropriate to the seen world" (118).

Although scientists have vigorously explored the "unseen world" of atomic physics, with its seemingly "mystical" migrations of atomic particles such as electrons, rational scientists have rarely been inclined to accept wholeheartedly the arguments of irrationality. These scientists quickly raise their eyebrows several degrees, and wonder what useful meaning if any is being offered when exponents of irrationalism submit arguments such as this by William Barrett in *Irrational Man:*

> The realization that all human truth must not only shine against an enveloping darkness, but that such truth is even shot through with its own darkness may be depressing, and not only to Utopians. But it has the virtue of restoring to man his sense of the primal mystery surrounding all things, a sense of mystery from which the glittering world of his technology estranges him, but without which he is not truly human (119).

In the modern era, man finds himself confronted by problems of his own making that seem almost cosmic in their proportions—the nuclear threat, terrestrial pollution through human by-products and wastes, and the unmanageable numbers of humanity itself. In this era, many people, not excluding scientists, have been attracted by various brands of religious mysticism. The established religions have not endorsed these mystic views, but it is clear that traditional religion, as well as rationalism, has failed to answer man's questions and console him in his fears. Thus, in their perplexity, people turn to seers, diviners, and self-anointed prophets.

In the 1960s, accompanying the Vietnam disaster, black riots, youth rebellion, assassinations, and growing violence, as well as economic crisis, interest spurted in such spiritual esoterica as horoscopes, galactic travel, extrasensory perception, and religious eccentricities. New cults sprang up and flourished. Interest in witchcraft spread widely. Devils came out of hiding to a new popularity. A poll taken in conjunction with the film "The Exorcist" revealed that nearly half the population believed in the existence of devils. Millions of books have been published and sold concerning ancient civilizations founded by space travellers, and thus offering fearful modern man "proof" that he is not alone in an unfriendly universe.

When Viking landed on Mars, that was the critical question: Life? Did earth life have a friend elsewhere, if only the simplest of protozoa? When Viking experiments produced no proof that there has ever been life on Mars, we felt lonelier than ever. This probably gave a spurt to interest in horoscopes, devils, and strangers from other stars.

In connection with horoscopes, most U.S. daily newspapers contain them, and even those who claim to be skeptical of such matters read the predictions for their birth signs and feel reassured when the news is good. The columns minutely examine the lives of all with respect to the positions of the planets at the instant they were born. It is difficult to understand the specific relationship between "fate" and planetary motion, but the emotional wish to believe in such a relationship is easily understandable. Belief in an ordained life provides a convenient excuse for failure. If the planets decide who we are and what we can accomplish, we have a pleasant invitation to inaction. If we find ourselves secure in an easy rut, we can stay there and be comfortable. The fault, dear Brutus, lies not with ourselves, but with our stars; so let's sleep late for another century or two.

On a somewhat more intellectual level are the books advancing the premises and promises of mysticism. These range from Norman Vincent Peale's *The Power of Positive Thinking,* with its superficial appetite for the obvious, to Barrett's *Irrational Man,* which considers the broad spectrum of western thought including the twentieth century influence of existentialism and technology.

> Technology in the twentieth century has taken such enormous strides beyond that of the nineteenth that it now bulks larger as an instrument of naked power than as an instrument for human well-being. Now that we have airplanes that fly faster than the sun, intercontinental missiles, space satellites, and above all atomic explosives, we are aware that technology itself has assumed a power to which politics in any traditional sense is subordinate (120).

Perhaps most popular and representative of contemporary books on mysticism are those of Carlos Castaneda (121). Castaneda beginning in 1968 with *The Teachings of Don Juan* has published four books of philosophical/fictional interviews with a 70-year-old Yaqui seer named Don Juan. Don Juan is introduced as a man enlightened with "true knowledge." Many of Don Juan's sayings remind us of the Zen masters, and the roots of his teachings seem to grow through the earth from the Far East to the American Southwest. Don Juan asks us to experience the world "without interruption" and to stop our "internal dialogue." Man is part of earth and nature. He learns only in the process of his journey to death. Death as man's goal preoccupies Castaneda, since man comes to knowledge as he dies. He seeks to reconcile the tragic and absurd fate of man, confronted in some existentialist writings (e.g., Camus and Sartre), with the mystical insights of Don Juan.

Originally an archeologist, Castaneda earned M.A. and Ph.D. degrees with his books, which aim at a marriage between existentialism and mysticism. His skill as a writer serves him well and often convinces a reader that the marriage has taken place and been happily consummated.

A difficulty with mysticism, however, is the fact that mystical experiences are highly personal. The persuasive mystic may convince others of his own experiences, but it is often difficult for others to share those experiences directly or concretely. The result is that even mystics seldom agree with mystics. Each account of mystical events is unique, from William Blake to Castaneda, though a case can be made that they have a common commitment to a mystical reality with ecstatic dimensions. Accounts of mystical matters tend to be as individual, chaotic, and beautiful as dreams; and about as useful, a rationalist might add. Obviously, there is much we still don't know about the mental and emotional nature of man. Human psychology may be an elderly art but it is still an infant science. Even rationalists readily concur that many human inspirations, intellectual breakthroughs, actions, and attitudes are not fully understood, and indeed do have a "mystical" aspect. Rationalists, however, simultaneously insist that the best prescription for mysticism is rationalism, and they shudder at the thought of establishing human policies on the basis of "mystical inspirations." They insist on cohesive, thorough, rational analysis of mystical ideas, to separate the useful and legitimate from the moonshine.

Perhaps even some agreement exists between mystics and rationalists on this point. The irony is that many who are "into mysticism" find it expedient in their practical affairs to rely on balance sheets, computers, and test tubes rather than the supernatural. Emily Dickinson wrote four lines of verse that seem to summarize the competition and partnership of mysticism and rationalism:

> Faith is a fine invention
> For gentlemen who see;
> But microscopes are prudent
> In an emergency! (122)

With better understanding of nature considered in their grasp, men thought themselves less in need of transcendental exploration. Today with that grasp less secure, and with some assumed answers becoming unsettled, the emotional need for supernatural consolations has become prevalent again. Religion as a platform for personal ethics enjoys resurgence because of the stormy times, socially and intellectually. As noted, mysticism has also attracted voyagers. Rationalists provide a persistent counterbalance, warning that to avoid capsizing and drowning in a mystical lagoon, prudent men will keep reason at the helm to steer and mind the rudder.

Modern society obviously relies on reason and rationalism to solve practical problems effectively. Even the mystic doesn't call another mystic but a plumber when he has trouble with the pipes. The consensus no doubt is that in the real world where the body (if not always the mind) has its permanent residence, mysticism has little if any utility in solving society's problems. That responsibility goes to the scientist and engineer equipped with facts that will stand up and figures that will add up. Of course, nothing stops the technologist, after his practical work, from going home, putting on sandals, and undertaking mystical explorations. Mystical speculations can have individual appeal. They have often led to poetry, as in the case of William Blake, and to spiritual experiences. But mysticism remains a recreation, a philosophical probing of the unknown, a technique of religious quest. Commonplace, everyday needs summon other techniques.

ARE NUMBERS EVIL?

Engineers and scientists habitually dealing with numbers do not find it easy to understand why some fear quantitative analysis. The fear arises, of course, from a humanistic wish to prevent numbers from taking the place of human beings, and to keep human needs from being lost in a maze of statistics.

The analyst often prepares a mathematical model of the system being studied. He adjusts the model so that certain inputs will produce known outputs. The model then can be used to predict system outputs with a variety of inputs and over extended intervals of time. Such model building is invaluable in the construction of physical systems. With humans involved, however, creating models and attempting reliable predictions becomes immensely complicated and potentially disastrous. Predictions are only as reliable as the stability and accuracy of the model. The introduction of human variables causes models to become imperfect representations; and dangers appear when planners treat a model and the humans involved as perfect replicas of reality.

Such models are frighteningly prevalent in the military, with entire armies and navies viewed somewhat as chess pieces on a chess board. By moving the pieces according to established rules, the "game" may be "won." In war, the players/planners sometimes forget that the "pieces" represent human beings. Read a history of any war and be struck by the inevitability with which numbers and statistics become the counters for playing a game, and how they rank as camouflage to conceal the dead. H. M. Tomlinson called military strategy "the metaphysics of disaster," and deplored its idiocy. War models also fail abysmally because human beings, with their infinite complexities, may not be willing or able to follow the established rules, with calamity as the outcome, totally removed from all the neat assumptions of the model makers. Vietnam is the most recent tragic example from a long catalog. The bombs fell, people died, the land was destroyed, terror ruled, and all the models were nonsense.

> 'Good morning; good morning!' the General said
> When we met him last week on our way to the Line.
> Now the soldiers he smiled at are most of 'em dead,
> And we're cursing his staff for incompetent swine.
> 'He's a cheery old card,' grunted Harry to Jack
> As they slogged up to Arras with rifle and pack.
> But he did for them both with his plan of attack (123).

In modern business as well as war, humanists find reason to criticize absorption in numbers and statistics. Efficiency experts study an operation, refine it with available technology, and determine to three decimal points how many are needed to run it. But human skills and energies don't always prove cooperative. Skills and energies differ widely among individuals. They even fluctuate seasonally and can be affected adversely by everything from an erratic sound to the baseball World Series.

A model that presumes to define explicitly how to maximize corporate profits can fall on its face and result in breakdown of the whole system. Tomlinson wrote that "facts and figures may be arranged, if you like, into something as pleasing as a floral design" (124). Then he described what had happened when this approach was applied rigorously in the British shipping industry by staffing a ship with "the smallest number of men . . . which could reach port if they all worked for their lives" (125). The result was that some ships did not reach port and many men were drowned. Tomlinson blamed an inordinate concern with numbers and statistics, and too little concern with human beings. "It seems hard to justify the health of the State," he wrote, "when its statistics require the faces of children to be pinched and hungry" (126).

Pushing our way through the emotional fogs, it becomes clear that numbers can be put to a bad and costly misuse, but they are not inherently bad and properly applied can be important. In the case of population growth, for instance, it is essential to consider numbers. If only a few million people lived in the U.S. as they did through most of the nineteenth century, there would be no population, pollution, or environmental problems on any scale approaching those of today. What is the optimal population for the U.S. of today with its current resource profile and geographic base? Constructing a model in which all available factors are applied does help to guide us in population and resource planning. The essential criterion, of course, in all model building, especially those models involving human variables, is to keep in mind the tentativeness of the model. In addition to human variables, those of nature must also be taken into account. During the last 50 years, man has learned that nature sets constraints. The environmentalist slogan "nature bats last" stresses the limitation of resources and the precariousness of life on earth. We must apply this knowledge in future model constructions and control measures. Otherwise the "numbers" may betray the future and turn good scientific intentions into misfortune.

INSTINCT VERSUS REASON

Many writers have considered the dichotomy between emotions and reason. Strident critics of technology such as Roszak and Laing openly oppose rationalization and opt for unanalyzed, direct experience. With this approach, planning is suspect, and reality is compressed to the here and now. The theory is that when we are in tune with nature, our bodies will move to an inner rhythm harmonizing with nature's rhythm. We shall lose all sense of strife and worry. Proponents of Zen Buddhism offer similar ideas. "Submerging into one's true self to find the all."

An assumption underlying the oneness-with-nature approach is that instincts are wiser than reason, and that primitive men with their largely instinctual lives had fewer "hangups" than modern men, and were therefore happier. All animals live purely by instinct, and thus presumably know the utter contentment of the present hour. The same perhaps with primitive, instinctive man. But is happiness a measurable or even recognizable event without the intellectual possibility of unhappiness?

Aside from the highly relevant questions concerning the value of mindless instinct in connection with happiness, we can question whether primitive men, who were not animals and thus not totally creatures of instinct, really had fewer hangups. Knowing little about the world around him, and with fewer explanations for visible phenomena, man could have been assailed by fears and more troubled than we now imagine. Few today would risk relying entirely on nature for their well-being when they observe what nature proceeds to do with any neglected patch of ground.

Also one could doubt if our various bodies would somehow contrive to shake harmoniously to the same inner rhythm. If not, confusion would be the by-product of this quest for human freedom, with each person shaking alone in a private solitude.

Considered over hundreds of centuries, nature seems remarkably bipartisan, as fond of insects and coyotes as man. She seems fondest of versatility and adaptability. If birds, insects, plants, and man can adjust to changing environments, nature tolerates them cheerfully, and they are given more times at bat. If not, out of the game they go, and nature doesn't seem to care or to look back. Placed within the framework of evolution, man is simply one of the more adaptable events in a long process. Does he now show signs of lessening adaptability? Has he trapped himself into dependence on external trappings, possessions, things, petroleum? Will man allow himself to be another casualty in nature's continuing experiment as was the dinosaur, who had a vastly longer run of it than man so far has known before it declined into extinction?

Writers such as Roszak attract many to a spiritual belief in nature without convincing them to rely on it. Usually they still expect science and rationalism to pay the rent and keep the roof from leaking. The celebration of inner rhythms and natural harmonies may, however, lessen respect for science and reason, just as children on occasion develop contempt for the work their parents must do to win them bread. This haughty attitude toward science and reason can open the door to an arrogant anti-intellectualism.

Caution is advised therefore. Unanalyzed emotions have led more than once to lynch mobs. Dependence on such emotions comes much too close for comfort to the "think with the gun" instincts of totalitarian regimes and exponents. There were millions in Nazi Germany listening to "inner rhythms" and when the Fuehrer spoke, those rhythms came into frightful harmony one with the other until they shook the world.

Lorenz seems to say that man's intelligence cannot compete with his instinctive, aggressive genes. Roszak similarly wants to disconnect rationality and rest in the soft arms of instinct.

REVOLT AGAINST REASON

Charles Frankel, Professor of Philosophy at Columbia University, examined irrationalism in a 1973 issue of *Science* (127). Frankel saw five fundamental propositions linking irrationalists:

1. The universe has two realms, appearance is one, reality the other. In the first is accident, uncertainty, alienation. In the second trouble disappears and harmony triumphs. Irrationalists favor the second.
2. People mistake appearance for reality because their ideas rest on social and political presuppositions. The method to achieve reality is to sweep the mind clean of rationality and to experience reality with the emotions.
3. War exists between the "cerebral" and the "emotional," between the "conscious" and the "intuitive," between the "empirical" and the "rhapsodic." Too much rationalization overweighs one side of the equation and dehumanizes man.
4. Distinction between subject and object is a reason to distrust science. It is man's fault if he is alienated from nature. Rationalism is to blame if we feel alienated from one another.
5. Human problems reduce to a loss of harmony between man and environment, between head and heart, between ideas and instincts. When harmony is regained, man lives free from unrest. He is no longer fallible or vulnerable. He can live in pleasant contemplation or mystic ecstasy.

Replying to these views, Frankel argues that science distinguishes better than irrationalists between appearance and reality. Science interprets the evidence of our senses and replaces conventional opinion with ideas supported by impersonal evidence (the explanation of planetary motion for example). Rational inquiry, rather than limiting the world of experience, enriches it. Explanations for geological formations, subatomic particles, the movement of continents, the distances of stars, electromagnetic waves, and countless other natural phenomena are conspicuous victories for reason and the rational mind. Such victories lend credence to the hope that in time man may also solve his other problems.

In such achievements, the optimism of science and scientists is fertilized for the future. These intellectual enrichments offer the attentive a spectrum of beauty and aesthetic reward.

Frankel points out that one cause for the assault on science is that science perceives nature as amoral and man simply a part of nature rather than the center of it. This decentralization of man proves indigestible for many irrationalists. It is too clinical, too cold, too objective. Nevertheless, the reality of the scientist can be tested by other investigators, while that of the irrationalist is established by *a priori* assumptions.

The irrationalist claims to find reality by floating on the sea of experience, imposing nothing. Frankel believes no inquiry can proceed without assumptions, probings, and inquiry into responses. Scientific assumptions are always tentative and subject to later correction. Reason allows beliefs to be collected, sorted, appraised, and ranked without recourse to force. Rationalism allows all ideas to be displayed, examined, and choices logically made. The irrationalist, believing that the needs of humans and the structure of the universe are symmetrically intertwined, asks the abandonment of reason and places all faith on human impulse and spontaneity. Frankel counters by noting that reason is as much a part of human nature as any other emotion. He suggests that we should carefully nurture it, since reason is weaker and needs greater attention than the blunter emotions. Frankel considers spontaneous impulses potentially and frequently disharmonious, with only reason available to play the harmonizer.

The irrationalist believes in a good universe, and accuses rational man of seeking to make it or portray it evil. Frankel calls this attitude a restatement of the myth of the Fall of Man. The irrationalist offers man the role of Eros, rather than that of Prometheus, and asks us to have faith that constraints do not exist for man in the universe of the instincts and emotions.

During periods of turmoil and drastic change, irrationalism offers an easier escape than the hardship of struggle. Struggle can be aimed at a specific goal, and thus success can be measured. But if success can be measured, so can failure. If reason can be avoided, so can failure, since only reason can identify failure.

Gerald Holton in *Daedalus* also wrote perceptively about the attack on rationality (128). Holton's point of departure is Reich's *The Greening of America* and its advocacy of heeding the instincts, obeying nature's rhythms, and being guided by the irrational. Holton shows how such ideas go back many centuries, hence his name "Dionysians" for the opponents of reason, and "Apollonians" for their opposites in the scientific camp.

Holton feels that much criticism of science rests on misinformation concerning how scientific work proceeds and on the way scientists report their work to the general public. This echoes concerns of C. P. Snow in *The Two Cultures* and other works. Holton notes that a scientific paper reveals little of what actually went on, the mistakes, wrong paths, errors, anxious and exhausting effort, and imaginative leaps. Scientific papers too often go out of their way to eliminate such human problems or concerns, and give the impression that research is cut and dried to a mathematical certainty. Holton feels "a deeper involvement of research scientists in discussions concerning their methods would surely improve the understanding of science."

Holton quotes Peter Medawar who said that "the process by which we come to form a hypothesis is not illogical but non-logical." Once an opinion is formed, however, it must be rationally examined for weaknesses and flaws. This is the scientific application of rationality, and it has equal relevance for other fields of human endeavor. Thus science both goes beyond logic and uses it. It goes beyond the rational and yet relies on reason to provide a foundation and a technique where all suppositions can be verified. When a scientific "leap" is made, much in the instinctive manner of mystics, rationalism must then step by step meticulously find out whether or not the truth has been reached or something other than the truth.

This process was described in exciting detail by James Watson in *The Double Helix,* which brilliantly shows how creative science really happens, not how critics think or pretend it happens (129). Watson's book concerning the laboratory search for the basic structure of life, is a genuine thriller, as improbable, human, and compelling, as any emotional voyage of the spirit by a practicing mystic. It helps prove that a revolt against reason implies a revolt against creativity of any sort, including artistic creativity (which is also innately logical and orderly).

Both artists and scientists find reason too useful for abandonment. It is improbable, of course, that instincts and emotions will replace reason. The "reasonable" assumption is that both reason and instinct will continue to be needed by man for survival. Technology, the same as reason, will not be abandoned either, and it seems clear that harnessing technology is a continuing challenge and obligation. In this crucial effort, all can contribute, irrationalists as well as rationalists. Indeed all will be required.

Technologists in particular may need to concentrate on defining man's goals. Perhaps the most valid criticism of technologists has been the charge that they don't know where they're going (like the flight captain, perhaps apocryphal, who announced to the passengers, "I have two pieces of news to report, one good, one bad. The bad news is, we're lost. The good news is, we're making really excellent time."). According to Holton, this problem can be due to the impersonal nature of scientific communications. Holton asks for an accommodation between the classically rationalistic and the sensualistic components of knowledge. This means that scientists possibly should tell all, or at least more in their writings, as did Watson in *The Double Helix.*

Technologists and artists have their work waiting for them. It is frighteningly evident that we are in trouble. If pollution doesn't get us, overpopulation might, or one of the many other spectres approaching man in his matchstick fort. Obviously dramatic political changes are needed to free technology for beneficial use. But what changes? How can they be made? As a survivor of World War I and with another World War approaching, H. M. Tomlinson decided that nationalism is the worst spectre of all, and the patriotic idea "My Country, Right or Wrong" the most dangerous idea afloat.

> Hadn't we better face it? The world will remain unsuited to a life of reason, with all the peaceful labor in it we want, while patriotism has a worse effect than opium. We must be now within a few years of the time...when patriotic propaganda will be regarded in the same way as traffic in noxious drugs. I suppose all of us have seen a man, otherwise sane, turned by a dose of patriotism into a rigid and hopeless fool quicker than any drug could do it (130).

There is work for technologists, and work for artists. Tomlinson over 40 years ago defined work that needed doing in the minds of men and among nations. The work still needs doing. Most of the world's great problems will not have technological answers. They require radical alterations in men's ideas and attitudes. But emphasizing the need for such alterations doesn't bring them about or even start the process of achieving them. We face serious situations that could overwhelm us. It appears unlikely that we shall have the option of retreat. Mars? Tomlinson's 1935 book is prophetic in its title, *Mars His Idiot.* It seems plain now that other planets, and possibly all of space, offer no easy options to man and no escape hatches. Can technology do more right here? Or rather, can man do more right here, with his instincts and his reason? Can rationality working in the area of human relations succeed as it has in some areas of nature? Time, if given a chance, will tell.

REFERENCES

111. Bane, G. Ray., "A Cabin in the Wilderness," *Defenders of Wildlife News*, March, 1973, pp. 194-195.

112. Hoffer, Eric, *The Ordeal of Change*, Sidgwick and Jackson, London, 1964, p. 2.

113. Russell, Bertrand, *Why I Am Not A Christian*, Simon and Schuster, New York, 1966.

114. Russell, Bertrand, *Unpopular Essays*, Simon and Schuster, New York, 1967, p. 9.

115. Oppenheimer, J. Robert, *Science and the Common Understanding*, Simon and Schuster, New York, 1966, p. 98.

116. Einstein, Albert, *Ideas and Opinions*, Bonanza Books, New York, 1954, p. 40.

117. Eddington, Arthur Stanley, *Science and the Unseen World*, The Macmillan Company, New York, 1929, p. 41.

118. *Ibid.*, p. 75.

119. Barrett, William, *Irrational Man*, Doubleday Anchor Books, New York, 1962, p. 275.

120. *Ibid.*, p. 201.

121. Oates, Joyce Carol, "Don Juan's Last Laugh," *Psychology Today*, September, 1974, pp. 10-12, 130.

122. Dickinson, Emily, *Selected Poems*, The Modern Library, New York, 1924, pp. 30-31.

123. Sassoon, Siegfried, *Other Men's Flowers*, Anthology of Poetry compiled by A. P. Wavell, Penguin Books Ltd., Harmondsworth, Middlesex, England, 1960, pp. 96-97.

124. Tomlinson, H. M., *The Foreshore of England*, Harper and Brothers, London, 1926, p. 35.

125. *Ibid.*, p. 52.

126. *Ibid.*, p. 69.

127. Frankel, Charles, "The Nature and Sources of Irrationalism," *Science*, Vol. 180, June 1973, pp. 927-931.

128. Holton, Gerald, "On Being Caught Between Dionysians and Apollonians," *Daedalus*, Summer 1974, American Academy of Arts and Sciences, Boston.

129. Watson, James D., *The Double Helix*, Signet Book, The New American Library, New York, 1968.

130. Tomlinson, H. M., *Mars His Idiot*, Harper and Brothers, New York, 1935, p. 193.

QUESTIONS FOR REFLECTION

1. Which should have the dominant voice in directing society: scientists and engineers or mystics?

2. In the interests of harmony and human progress, should mystics and rationalists, in the manner of the lions and the lambs, seek to lie down together? What would be accomplished?

3. Should scientific investigations not occur in certain areas? Which areas? Why?

4. The availability of atomic energy has raised questions and problems. Would we be better off if the release of atomic energy had not been accomplished?

5. How and by whom should technological developments be controlled? Can technological developments be controlled? What are the consequences?

6. To what extent does history encourage optimism that man can improve his status both physically and mentally?

7. What is the optimism of science and how does it serve man's future?

8. H. M. Tomlinson wrote that "change is not progress." Define change. Define progress. What do they share? How are they different?

9. What *are* man's goals, defined in terms of his present ambitions and behavior? What should his goals be?

10. How is reality likely to alter man's future goals?

10. Sources of Values

"As we see, what decides the purpose of life is simply the programme of the pleasure principle."

Sigmund Freud
Civilization and its Discontents

"It was by learning that we ceased to be animals and made ourselves into men."

Gilbert Highet
Man's Unconquerable Mind

SYSTEMATIC VALUES

Freud wrote *Civilization and its Discontents* in the summer of 1929. The title he originally intended for the work was *Das Unglück in der Kultur* (The Misfortunes of Civilization). But the contents focus instead on the disenchantments and discomforts that are man's lot because of the eternal conflict between his instincts and the requirements and social regulations of civilization. The first sentence of Freud's book places us at the heart of our own concern:

> It is impossible to escape the impression that people commonly use false standards of measurement—that they seek power, success and wealth for themselves and admire them in others, and that they underestimate what is of true value in life (131).

Heredity and environment shape individual reactions to situations in life through the value systems they jointly influence man to develop and adopt. Many factors contribute to the value systems that each man carries with him. Thinking processes and habits are formed by parents, peers, teachers, public figures we admire, writers we read, actors and actresses, achievers we seek to emulate. Thus as the individual matures, consciously or unconsciously, he develops his own individual value system. The process is inescapable and inevitable, but often the resulting system appears sketchy, contradictory, with fuzzy boundaries and a lack of precise definitions. Individual value systems may prove to be mirrors of the society, but they sometimes reflect the worst values of the society rather than the best. The possessors of flawed or unexamined value systems are sometimes shocked into a critical consideration of their values by personal or social crisis, and they may be dismayed at what they learn.

The happiest individuals are those who possess distinct value systems that have been consciously and methodically appraised, and brought into harmonious relationship with their lives. Such systems, openly arrived at and actively maintained, are easier to adjust and alter when necessary. Ideally, of course, values provide the strong foundation stones on which we build our lives, and when they are solidly fixed in place, they are moved or changed only after long contemplation.

The ideal, however, has rarely been attained. In our technological civilizations, values have too often been built on shifting sands. They have washed away in the first crisis. The construction of new and reliable value systems has long been our most important task, with innumerable criteria to consider from man's view of nature to his ideologies and philosophies.

NATURE AS ANTAGONIST

According to Gilbert Highet, man's "essential history is the story of our learning and thinking." The first stage in the transition from animal to man was completed with the development of the ability to learn and remember.

> It was then, far back in the warm jungles, that somehow, cell by cell and reflex by reflex, that the wonderful human brain was formed, and with it our two other human powers—the devices by which even if the world fell into ruins, we could still rebuild it—our fantastically intricate speech, and our ingenious adaptable hands (132).

Early man with that developing brain probably would have groaned with weariness if Highet's impassioned tribute had been explained to him. For all his wonderful brain, life was an enormous challenge and hardship. Obtaining food was difficult and dangerous. Nature, far from being a friend, seemed inexorably hostile. This was true for hundreds of thousands of years during the tedious evolution of social organization and civilization.

In modern times men have nervously claimed certain successes in "controlling" and "altering" nature. The nervousness arises because such success cannot be taken for granted. A river "controlled" for normal circumstances, may be turned to a raging and lethal flood by abnormal circumstances. "Act of God" some might say or "Nature running amok," but whatever words are applied, it is apparent that nature even when subjugated and harnessed can and eventually will burst all harnesses and cataclysmically run free.

Nature sooner or later tends to show a hostile face to man. The farmer works long hours over many months to see his work destroyed in any one of half a dozen natural ways: insects, cold, heat, drought, hail, flood. Sailors watch the sea turn glassy and the needle on their barometer move lower, and they get ready for fresh trouble from nature. On frontiers of the past, acquiring the necessities of food, clothing, and shelter was difficult, and men were esteemed by their neighbors when they were especially adept or could find more efficient means for obtaining these requirements. The Antarctic and space are the chief frontiers today, and on these frontiers as on those of the past, explorers survive by confronting nature, the antagonist, and displaying the tenacity and the skill required to win.

In the development of North America, Americans saw nature as an antagonist perhaps longer and more actively than was necessary. "Conquer the wilderness" was the rallying cry when the continent was vast and untamed. Gradually the cry turned to "Exploit the land" as there was less and less wilderness to conquer. Even today, though technology has clipped some of nature's fangs and curbed antagonism, an apparent human impulse for plunder and exploit continues. Technologists have obeyed the general trend to build and develop with minimum concern for the environmental consequences. Today, however, thanks in part to the success of technology in giving men the leisure and security to do other things than struggle with antagonistic nature for survival, a growing number of voices cry, not in the wilderness, but for survival of the wilderness. They suggest that there are healing forces in nature men need and should preserve. Yet the general attitude toward nature remains one of antagonism. When the chips are down, people choose houses and automobiles over wilderness. Technology has given them highly competent means for dealing with wilderness and reducing it to suburbia. If "reducing" has a denigrating sound, it can optionally be replaced by the word "promoting."

So nature remains the antagonist to be confronted and compelled to submit. But the antagonist is never perfectly conquered. Dandelions and other weeds sprout defiantly in the best manicured suburban lawns. Unfriendly birds do unfriendly things to fresh paint. And often in the night, houses built near the San Andreas Fault in California rattle suggestively as a warning to the sleepers within that nature is not asleep, and has nasty schemes in mind for a future time yet to be determined.

NATURE AS BENIGN

Some philosophies and religions, particularly those of the East, see nature as a benign force and friendly presence. In such religions, man strives for identification with nature and obedience to nature's rules. In western culture, pantheism repeats these Eastern sentiments by identifying God with everything in nature. Originally conceived by Spinoza as a means of uniting mind, matter, soul, and body, pantheism appealed to romantic poets Wordsworth and Shelley, and was made an influential factor in American thought by the writings of Emerson and

Thoreau. Pantheism traditionally has attracted intellectuals who cannot accept orthodox creeds.

The national religion of India, Hinduism, emphasizes the benignity of nature, and stresses the importance of man's collaboration. Thus, cows, monkeys, and animals in general are sacred (because they possess souls destined for eventual migration to higher life forms). Being sacred, they cannot be harmed or used as food. To a people often on the verge of starvation, it seems a costly indulgence to live in this manner. By contrast, Communist China adopted western ideas of progress and has remarkably reduced starvation in one generation, to the discomfiture of India's friends who see less progress there.

In the U.S., American Indians often adopted the benign approach to nature. Navajos, in particular, developed an unusually tolerant, as well as stoic attitude toward nature. This posture served them well in their hostile environment. But it did not, of course, prevent them from being harassed, conquered, and economically subjugated by the westward hordes of Europeans who saw nature as an antagonist to be overcome and Navajos as part of that nature.

Modern critics of technology tend to adopt the benign approach to nature. As we have seen, they assume a pantheistic attitude which removes man from the center of the universe and makes him merely one of many living species. Properly attuned to nature, man's inner needs will be developed and satisfied. Benign nature will generously deliver the other necessities of life. But except in a few cases, enthusiasts for this point of view have shown a singular reluctance to quit their jobs or cancel their welfare checks and trust themselves totally to nature. In practical matters, the benign approach must seem to them a little reckless; yet some have found peace of mind, relaxation, and new vistas for contemplation through hikes in isolated areas and occasional opportunities to experience wilderness. Renewed vigor is one conspicuous benefit of such outings. Benign meetings with nature are clearly spiritually medicinal and curative for urban and suburban man who can so easily lose identity with nature. Permanent residence on the side of a mountain in the midst of towering pines will not be an available choice for most men, but keeping such places available for all men to experience has to be considered high on man's priority list of things to do.

NATURE AS INDIFFERENT

In some religious teachings, man has an extremely eminent position; but knowledge supplied by the facts of evolution and astronomy tend to flatter man less in one sense, and yet to enlarge him in another. Considering the billions of suns in the universe that dwarf our own and the millions of lives that arose on this tiny planet of one insignificant sun and then fell by the wayside, man appears as one of countless experiments performed by nature. All of man's history encompasses but a tiny fraction of the four to five billion years of earth's existence. Considering such facts, we might view nature as neither hostile nor benign, but indifferent. Man is simply a tiny part of nature. If he chooses to make himself extinct, nature will not mourn but will move on to another experiment, perhaps one that works out better or is less destructive and more fun.

Nature's indifference does not mean that man can do as he pleases without paying a fee. Nature conscientiously collects for each of man's conquests. Thus, for the use of oil, man pays by enduring smog. For tilling the soil, he accepts erosion. For the use of insecticides, he pays with the lives of other species and with his own health.

Man has plundered the earth unconscionably and displayed virtuosity in despoiling his own habitation, but it would be useless to follow some critics in assailing man and arguing that the earth would be better off without the human race. Only man has developed the values necessary for observing and evaluating the facts of plunder and spoil. Only man has the mental subtlety for criticizing himself and deliberately choosing to alter his established habits for the sake of the future. A world without man would be a world entirely of instinct and habit. It would be a world without values.

Steven Weinburg in *Daedalus* stresses the difference between nature and man when he writes, "The laws of nature are as impersonal and free of human values as the rules of arithmetic. We didn't want it to come out that way, but it did . . . Having committed ourselves to the scientific standard of truth, we have thus been forced . . . away from the rhapsodic sensibility" (133).

In the same issue of *Daedalus,* Theodore Roszak counters with insistence that a monster of meaninglessness has been produced by science in the modern world. "There is no experience of the divine, only the experience of man's infinite aloneness." Roszak asks if we can be certain that what science gives us is genuinely knowledge. He chooses as his metaphor for technological endeavor and achievement—Frankenstein's monster. Frankenstein's creation, though benevolently intended, emerged as a monster, by Roszak's measure, because it lacked a soul. Roszak thinks that science methodically turns man into a machine and removes his soul.

Roszak sees science slicing nature into minute parts, investigating each intensely. The results include reams of information and "reductive knowledge." But the results do not include intimate knowledge of the whole or of the "essential quality." Roszak deplores that science must stop short of the "religious experience." He regrets science's incompetence to contain the "aesthetic shape of the world." Concerned that power (science technology) without spiritual intelligence becomes a monster or a producer of monsters, he expresses the view of the "other culture" ably in this essay (134). He also makes clear the developing public disenchantment with science. The contributions of technology are still appreciated but they are now hedged about by suspicion. Technology has not solved man's problems. Indeed, it may have increased his uncertainties by bringing his ancient vision of himself into question, by regimenting his life with imperious machinery, and by forcing him to examine critically his own values. It is apparent that the truths of science can serve to weaken or even effectively condemn existing value systems.

Roszak warns about the dangers of scientific endeavors detached from human need. He wants to protect the soul of man. Yet others, equally sincere, wonder if commitment to the supernatural is not a hindrance to man and a weight more inhibiting than any burdens imposed by technology. Barrows Dunham, for instance, speculated that if men were convinced they possess only the here and now, and have only the opportunity of their present lives, if they might not "strive to make that chance successful, that here and now bright with all possible bliss." Dunham thought this recognition of a common fate might help men identify with a common purpose and serve to "end the schisms which divide them" (135).

In a sense, scientific knowledge has reduced man's importance in his own eyes by showing him his tiny place in the universal sea of stars. Yet man has proved himself able to reach out with his mind to the farthest horizons of that sea. Only man has equipped himself to count the galaxies and even to visit them with his instruments and restless speculations. Only man has asserted himself on both land and sea with both finesse and brutality. Only man has acquired the strange knack of beautifying as well as poisoning. And only man has turned dead leaves into power and loneliness into poetry, establishing competitive values of technology and the spirit, equally valid, equally necessary.

Nature may be indifferent, but man is not. His own driven nature gives him a certain tenuous dominion over earth's continents and oceans, but as he floats on earth's oceans and gazes outward into the oceans of the universe, he senses his terrible responsibility, and his capacity to scale heights but also to plummet into depths. And sometimes, dismissing the artificial and transitory concerns that besiege most of us most of the time, man's responsibility burns in him like a fever. He has even managed to view his planet from space and to marvel at the phenomenon of life upon its surface. And maybe, just for a moment, his values include some transient impulse to protect and preserve what he views.

IS MAN'S NATURE AGGRESSIVE?

Is man fundamentally aggressive or cooperative? This question is important, and the answer is still being sought. If man has an inherently warlike and destructive nature, then it may be impossible for humans to reach a plateau of peace where men can work harmoniously together.

Robert Ardrey wrote *African Genesis* (136) and *The Territorial Imperative* (137) after interviewing paleo-anthropologist Raymond Dart. Dart's theory that an early race of men had used bones and rocks to kill and had been flesh eaters was elaborated by Ardrey into the conclusion that early man was a predator and natural killer. His thesis is that man's first tools were weapons and that the development of superior weapons has been his chief occupation. Ardrey implies that war has been an essential activity of man, and without it he might decline. Without new weapons to develop, man's future would be uncertain, like a tippler outside a saloon on election day. Yet Ardrey sees the peril. As his weapons have reached the atomic threshold, man's threat to himself has reached the threshold of obliteration. Ardrey hopes the human mind can save itself by making an alliance with the instincts. He believes this possible though he thinks that instinct will always win a contest between man's instincts and his reason. In this thesis, Ardrey joins the critics of technology who consider instincts more to be trusted than reason. Yet, if Ardrey's reading of man's character is correct, it would seem that reason should be sent urgent cries for help.

A scientist studying animal behavior, Konrad Lorenz, has stronger credentials than Ardrey to interpret scientific evidence concerning human behavior. Lorenz bases his theories on observations made of animals defending territory (138). He determined that many animals establish territorial claims and attempt to exclude others of the same species. He contends from his observations that aggression, defined as the fighting instinct directed against members of the same species, is crucial in the natural selection process. In animals, aggression seldom results in death, but Lorenz suggests that in man the mechanisms that inhibit fighting in animals may fail due to the character of modern weapons and human social forms. Lorenz thinks society might sublimate aggression through better understanding and alternate methods for directing such tendencies into safer channels. He supports some of Ardrey's contentions concerning man the aggressor. Together they provide a rather grim composite picture of human hostility.

A number of scientists disagree with such conclusions (139). For one thing, these investigators question the extrapolation of animal studies in this manner to human beings. Though animal studies usefully relate to man in connection with anatomical similarities, psychological and social relationships are much more questionable. Many doubt that man possesses instincts comparable to the instincts of animals. They argue that man's behavior is socially learned. Scientific issue has been taken with the methods used by Lorenz and the startling extrapolations made by Ardrey. For instance, we know that the simians, with which man seems most closely related, normally exhibit unaggressive behavior. Köhler in his classic *The Mentality of Apes* (140) describes the clannish friendliness, the quick tolerance of outsiders, the eagerness to forgive and be forgiven. He tells how stones would be thrown cautiously at one of the apes, Tschego, as a form of punishment. ". . . Tschego, as soon as one of us was scolding her and at the same time picking up a stone to throw, would run to that person, take hold of his hand and hold it tight, without showing any special excitement. Then she would coolly take the stone out of his hand, throw it on the ground, and walk quietly away" (141).

In addition to scientific doubts, when we examine the social utility of the human aggression theory, doubts multiply. Lorenz and Ardrey have become highly popular for their contemporary writings, but there were earlier tracts exploring similar ground. Herbert Spencer, for one, attempted to apply Darwin's "survival of the fittest" theme to human social and economic realms. The argument that those who were most successful in Victorian England were the "fittest" appealed, of course, to those at the top; but the argument lacked scientific substance. Spencer's book, *Social Statics,* actually appeared before Darwin's *The Origin of Species.* Though somewhat deficient in logic, *Social Statics* did express views compatible with modern efforts to identify and establish value systems allowing as many as possible to achieve "fitness for survival." "Education has for its object the formation of character," wrote Spencer. "No one can be perfectly free till all are free; no one can be perfectly moral till all are moral; no one can be perfectly happy till all are happy," he said. One can imagine the Victorian establishment dismissing such sentimental whimsies and locking firmly to "survival of the fittest" as justification for protective aggressiveness to remain at the top of the heap.

The aggression theory, despite rejection by some scientists, has continued to our own time, erupting periodically in books such as Ardrey's. The popularity of the theory is not mysterious. It serves as an obvious comfort to the "have-gots" who are encouraged to do whatever is necessary to preserve what they have. At the same time the theory reprieves man from moral responsibility and dramatizes his aggressive behavior as an established part of nature. However, since the theory has not been proved, and since the behaviors of animals or primitive men are not reliable guides to the behavior of modern man, the current trend is to be skeptical of the aggression theory and to emphasize the need for man to improve his social relations with a lessening of aggressive behavior, and to offer the reasonable hope that both are possible.

The theory of inherent human aggressiveness becomes involved in such perennial questions as inheritance versus learning and genetic influence versus environmental influence. How much of a man's behavior does he inherit from his parents and ancestors? How much is derived from his surroundings? These questions have produced intense arguments and much research (142). Some findings have been highly controversial and passionately challenged. William Shockley, for example, claims that the known difference in IQ test scores between blacks and whites results from the inferior genes of the blacks. As remedies, he proposes payments to blacks for not raising children and encouraging an acceleration of racial intermarriage. Shockley's critics emphasize that blacks and whites come from different environments, that IQ tests stress verbal aptitude, and that chronically weak verbal skills too often characterize black ghetto educational results. The controversy goes on, and it tends to shed more emotional heat than scientific light.

Studies on twins have indicated that environment affects intelligence, usually as measured by the IQ test, but not as strongly as heredity (143). Clearly both environment and heredity operate on each individual. Equally clearly, insufficient knowledge exists concerning the operation. Improved environmental factors have produced markedly beneficial results. Genetic factors are also conspicuously important. Both environmental and genetic control are increasingly within reach of modern technology, but caution must be applied, particularly in genetic control. Doctor Frankenstein's experience warns us to go slow, to make certain of our goals, and to beware of irreversible consequences.

A more fruitful pursuit in the immediate future would seem to be further research into the nature of man with the direct scientific goal of man learning about man as intimately and comprehensively as possible.

TYPES OF PHILOSOPHY

Idealism

Plato was the historical founder of idealism; and it was developed by a number of later philosophers, such as Locke and Berkeley, until it reached a zenith in the work of Hegel. Idealism holds that the Absolute or the Idea contains the universe. The earth and its multitudes are imperfectly observed reflections of the perfect Idea. Some philosophers have maintained that the source of the progenitive Idea is God, but this is not indispensable to the system. The ideal system of the universe and life operates eternally, perfectly. We may not understand, but this is due to our inability to comprehend all that the Idea encompasses.

In connection with values, if reality is permanent, then ideals also are permanent. Values are derived from the authority of permanent ideals and themselves in turn have permanence. Because of this permanence, values do not change, unless the absolute ideal requires it, and then they change slowly. In practical terms, the absolute authority of the ideal may be delegated in part to individuals or groups capable of attracting followers, and these individuals or groups may then for a time become arbiters of values. Founders of religions such as Jesus, Mohammed, Buddha, are examples of such authorities. Often several authorities exist simultaneously. This can lead to confusion, as in politics, or lead to war, as in the clash of ideologies. Authorities always resist change, hence the potential for disaster when comparably strong authorities compete.

The authoritarian promulgation of values, evolved from the philosophical tradition of idealism, probably constitutes the chief means by which values are established today. The state, church, parents, and other authority figures issue edicts to guide the individual. Many such edicts are derived from past experience. They contain prudent warnings that certain things are "bad" or taboo, such as stealing, adultery, incest, sticking one's finger in the fire, and voting for Democrats (or Republicans). Instincts may also be the source of value obligations for both men and animals. Köhler noted the existence of value taboos in chimpanzees during his observations. The animals appeared frightened and guilty when caught performing acts that were instinctively forbidden (144).

Among men, problems arise from the fact that values established by authoritarian edict tend toward rigidity. Rules and values that may once have been useful become sacrosanct and remain untouchable long after their validity has gone. To question such values, even when they are of "the Emperor wears no clothes" variety, labels one an apostate and heretic. If the value requires a fully clothed Emperor, then even when conspicuously bare, he is, to the obedient, fully garmented. However, when edicts become too antiquated, they sometimes wither into disuse as citizens pay them nothing but lip service. The withering process tends to be drawn out and tediously slow.

Stability is vital in value systems to prevent chaos, but "culture lag" is a major defect of the authoritarian system in a rapidly changing environment. Rules, valuable in an undeveloped continent with a moving frontier, may not serve well in urban, industrialized societies. Guns provide an example. On the American frontier, it was taken for granted that a man dressed for the outdoors would wear sidearms and perhaps carry a rifle. This rule seems less relevant today for commuters driving into San Francisco for a day at the Stock Exchange or account executives in Connecticut catching the New Haven for a ride to Grand Central and a session of struggle on Madison Avenue.

The well-dressed do not require sidearms on Madison Avenue, New York City; but Madison Avenue has certainly become one of the reigning authorities with power to establish values in contemporary society. Advertising clamorously asserts that human odors are uncouth, that these shoes are fashionable, that my idea is respectable while yours is dangerous. Advertising has become impressively, and rather alarmingly adroit at persuading, converting, and often proving for a fee that the worse is actually the better way. By holding certain things attractively up to the light and cleverly shadowing others, advertising has made a lucrative business out of manufacturing and channeling human values.

If advertising and related arts of persuasion conflicted with authoritarian values, confusion and conflict would result. But advertising has largely made itself a vigorous arm of authority, praising what authority wants praised, condemning what it wants condemned.

Students of value systems expounded by national states and powerful organizations, such as churches, may detect a lack of universality in the edicts delivered. Contradictions and diversity are particularly prevalent in connection with political and economic values, less so in connection with moral principles. This may be true because economic values concern money, political values concern power, while moral principles simply define modes of behavior established by custom and experience over many years. Both dictatorships and democracies find it expedient to convince citizens that honesty is the best policy, that live and let live is neighborly, that obeying most of the Ten Commandments helps promote social order.

Pragmatism

This philosophical discipline developed chiefly in nineteenth century America. Its leading exponent, William James, sought mainly to slay the dragon of the "absolute" or to allay its influence and curb its sovereignty. James stressed facts, action, and the effects of power rather than abstractions, fixed principles, and semantic solutions, He protested dogma, artificiality, and unchallengeable finality in truth. He wanted the "universe without a lid" and sought to remove all the traditional lids he could (145).

The U.S. with its open frontier was ideal ground for a philosophy inclined to judge an action good if it succeeded, and an effort appropriate if it worked. During America's era of rampant economic expansion, almost any action brought results. Thus pragmatism seemed just what the American pioneer, tycoon, and advocate of Manifest Destiny ordered. When two competing actions succeeded, efficiency was the measurement to decide the competition. Difficulties appeared only when there were two equally valid choices.

As an example, James considered the option between two statements: "There is a God." "There is not a God." What is the pragmatic choice? James picked the first, on the grounds that it has an equal chance of being true. If it is true, the believer receives all the benefits and satisfactions of belief. If it is not true, the believer is no worse off, and he escapes the unpleasant consequences of choosing the second if it turns out there is a God with ardent distaste for being disbelieved.

Pragmatism, in other words, does its utmost to choose and bet on the right horse. It is less a philosophy perhaps than a method of expediency. Not surprisingly, pragmatism has established itself as the philosophical method of many engineers and scientists. The ubiquitous question in technology concerning every proposal is this: "Will it work?" Variations of the question include: "Is it pragmatic?" "Will it float?" "Will it get off the ground?" If the answer is "yes,"build it. Questions about the purpose and possible consequences of a proposal have often received little or no critical analysis. The results of such a philosophy, plus limited vision, can be seen in atmospheric pollution, dump heaps, dead rivers, and sick oceans. It has become increasingly clear that "will it work?" poses only one of several questions that must be answered conscientiously if man achieves a responsible technology and survives his own gadgets. Also man needs sharp vision of his goals and understanding of his basic purposes. The philosophy of "whatever succeeds is right" though efficient in practical terms and in choosing what will work, is insufficient to sharpen man's vision and to shape moral values for himself and his technology.

Utilitarianism

English philosopher Jeremy Bentham was the leading advocate in the late eighteenth and early nineteenth centuries of an ethical doctrine that came to be known as "Utilitarianism." The name was popularized by Bentham's leading successor, John Stuart Mill, in his essay *Utilitarianism,* which further developed and explained the main tenets of this moral philosophy (146).

Utilitarianism commonly has been identified with the point of view that legislation and conduct should strive to promote "the greatest happiness of the greatest number." "Utility" was stressed as the surest means of accomplishing this goal, and exponents of utilitarianism sought to reconcile the broad goal with individualism. Bentham held that "pain and pleasure" govern mankind, and he argued that individual effort expressed through the principle of utility would maximize pleasure. His definition of the principle of utility makes a frank appeal to the cause of self-interest: ". . . that property in any object whereby it tends to produce pleasure, good or happiness, or to prevent the happening of mischief, pain, evil or unhappiness to the party whose interest is considered." The pursuit of pleasure was seen as the natural expression of morality. This has been interpreted various ways, as an invitation to selfishness, and also as an obligation to seek the greatest happiness for others. Mill refined the doctrine by placing stress on qualitative rather than simply quantitative pleasure, and Herbert Spencer adapted the philosophy to his scheme of social evolution.

Utilitarianism had strong influence on the nineteenth century. The roots of pragmatism clearly have some rootlets planted in the doctrine. It affected other philosophies and philosophers leading to modern times.

To measure happiness, Jeremy Bentham devised a mechanical grading system that critics of utilitarianism have challenged. Bentham proposed a "calculus" which individuals could use to decide on a course of action in a given situation. Mill's *Utilitarianism* explains the procedure. It consists of assigning number weights to various aspects of a proposed course of action, adding to the positive and negative sides, and reaching a decision on the basis of the greater total.

The technique could be applied to any decision-making situation, from the grand to the trivial. Thus, if the question were, "should we visit Aunt Helga on Saturday?" the following table might be constructed:

"Shall we visit Aunt Helga Saturday?"

POSITIVE		NEGATIVE	
Getting out of the rut	10	Automobile expense	15
Fresh air and sun	15	Missing the TV spectacular	10
Chance to shop on trip	5	Aunt Helga's mischievous son	10
Winning favor with rich Aunt Helga	25	Ruining the week's diet	5
		Traffic jams	10
	55		50

Based on this table, Aunt Helga should expect us as guests on Saturday.

The similarity of Bentham's "calculus" to a related scheme devised by Benjamin Franklin much earlier is a curiosity worthy of note. Franklin as a representative of the American Colonies in London was known and admired by Bentham. In 1772, Franklin communicated what he described as a new kind of "mathematics" with the title "Prudential Algebra" to British scientist Joseph Priestley. Prudential Algebra like Bentham's calculus involved making decisions on the basis of pro and con reasons why a certain action should or should not be taken. Franklin described entering the reasons for or against, much in the manner of the trip to Aunt Helga's, and after this, he wrote, "When I have thus got them all together in one view, I endeavour to estimate their respective weights. I have found great advantage from this kind of equation, in what may be called *moral* or *prudential algebra*," concluded Franklin (147).

Whether or not it is called a method of utilitarianism, this technique of decision-making is still widely used, if not as formally as it was by Bentham and Franklin. In the 1976 U.S. presidential debates, for instance, between candidates for the U.S. Presidency, one candidate described his method of judging a complicated tax bill. He would weigh the good versus the bad, and see which dominated.

Objections have been made to this aspect of utilitarianism on obvious grounds. The first involves the inexactness and difficulty of ascertaining all the criteria to be considered on the "Positive" and "Negative" side of the equation. Is it ever possible to remember everything that should be taken into account when contemplating a visit to Aunt Helga's? Probably not. The second objection concerns the arbitrariness of the weights assigned to the various factors. Even when exact numbers are not used (this was Franklin's version), the method has faults, especially when claimed as a definitive way to reach important decisions. Nevertheless, most of us automatically use Jeremy Bentham's "Calculus" or Franklin's "Prudential Algebra" in daily decision-making whether we realize it or not. If the method is not exact, it is a method, and vastly superior to flipping a coin. In the U.S., the method has been codified and widely used as the "impact statement" required for new constructions. Thus, expensive decisions are regularly made on matters of great importance using essentially this utilitarian approach.

Logical Positivism

Positivism began in the early nineteenth century when French philosopher August Comte sought to apply scientific method to philosophy and to social reform. Science then was struggling to emerge from childhood and adolescence into maturity. Comte, like the optimistic philosophers of the eighteenth century, looked to science to enrich and liberate mankind. Comte's social thesis was embodied in the phrase *"vivre pour autrui,"* (to live for the sake of others), which he saw as the mission of science.

Positivism influenced John Stuart Mill and other leading philosophers of the nineteenth century. Brought into our own century through the work of G. E. Moore, Ludwig Wittgenstein, Bertrand Russell, and others, positivism today competes with existentialism in modern thought. Sometimes called analytic philosophy, it has attempted to be scientific by subjecting arguments to precise analysis. The philosophy is anti-authoritarian. Suspicious of conceptual thinking, positivists want to replace it with a mobilization of facts. Add up all the facts, they argue, fully understand nature. Behaviorists, who explain science exclusively on the basis of specific operations and measurable results, are legitimate offspring of positivism.

Positivism is broader than pragmatism, but in its results, it seeks to be pragmatic. Ludwig Wittgenstein's answer to the question "What is your aim in philosophy?" was both pragmatic and compassionate: "To show the fly the way out of the fly-bottle." Positivists ardently concern themselves with the precise use and meaning of words. In the twentieth century, positivism found a major expansion through the influential school of linguistic analysis that developed at Oxford in England. One Oxford Professor, Richard Hare, told journalist Ved Mehta, who was interviewing British philosophers for *The New Yorker,* that "careful attention to language is, I think, the best way not to solve problems but to understand them. That is what, as philosophers, we are mainly concerned with." Professor Hare submitted that, "The thing wrong with the Existentialists and the other Continental philosophers is that they haven't had their nose rubbed in the necessity of saying exactly what they mean" (148).

From the standpoint of values, positivism asserts that concepts are inaccessible to empirical tests. Abstract words such as "good" or "bad" positivism considers merely expressions of personal likes or dislikes. G. E. Moore, Wittgenstein's teacher, claimed that "good" could not be defined, since the word itself forms a basic term that cannot be resolved further. Ved Mehta after immersing himself in British philosophy wrote of this approach:

> The logical positivism of the thirities, I learned, was a sceptical movement. It claimed that any statement that could not be *verified* by sense experience was meaningless. Thus, all statements about God, all statements about morality, all value judgments in art were logically absurd. For example, 'Murder is wrong' could only mean, at best, 'I disapprove of murder' or, still more precisely, 'Murder! Ugh!' What made a statement like 'There is a dog in my neighbour's garden' meaningful was that I could *verify* it. If I went into the garden, I could see the dog, beat it with a stick, get bitten, hear it bark, and watch it chew on an old bone (149).

Positivism serves the cause of exactness, and it has earned commendation for discarding superstition and fuzzy thinking. Yet, like pragmatism, it focuses more intently on method than on ethics. Pragmatism is a method of determining what will succeed, while positivism on a somewhat higher plane gives a method of determining what is fact. But positivism as a guide to values is no better than pragmatism. Positivism may assist with enormous precision in defining and understanding the matter "of the dog in the garden," but it offers little help with the open-ended and difficult questions, such as "what do we do next?" and "where do we go from here?" and "why?" Positivism is found wanting whenever we require value judgments. Scientists act on the assumption that any true statement is worthy of belief, which means that facts and ethics run in tandem. The logical positivist dismisses the ethical question as one not subject to verification.

Barrows Dunham has pointed out that the current popularity of positivism is not based on evidence but on the inclinations of contemporary society.

> Is this not a precise picture, a terrifyingly accurate picture, of our own commercial society? We know the vast physics necessary to control atomic energy; we know in lesser degree the biology needed to keep people in health. But we have not the social technology which would make these other sciences a blessing, and, our thoughts being more and more bent upon destruction, we have hardly any morality at all. Much physics, little sociology, and no morals: this is positivism, and this is what we are (150).

Positivism relies on physical science but wastes no time on "logical absurdities" (i.e., ethics). Yet science itself is inseparable from ethical considerations. Value judgments are made with every step in science. The scientist must choose the direction of his efforts. He must decide how observations should be made and with what instruments. He must trust the published results of other scientists. Finally, he must decide what conclusions are warranted, and often he has a responsibility to decide what uses should be made of them. Science has sometimes tried to ignore moral considerations, and has even used the philosophy of positivism as justification. But the effort never convincingly or consistently succeeds. Though positivism has joined pragmatism as a cornerstone of American social philosophy, it has never emerged as a true philosophy for the broader needs of humanity.

Existentialism

Existentialism shares the stage with positivism as the dominant modern philosophies. Though elusive of definition, existentialism can be identified to an extent as the antithesis of positivism. Where positivism is scholastic, analytical, cynical, and fact-oriented, existentialism is intuitive, mystical, passionate, and paradoxical.

Reading the classics of existentialism, by such thinkers as Kierkegaard, Dostoyevsky, Jaspers, Nietzsche, Heidegger, Camus, and Sartre, one wonders how and why they are grouped together since they seem to have so little in common. Kierkegaard wanted a "razor-edge decision" of the free will for a revival of Christianity based on faith (the opposite of sin). To Kierkegaard "despair is sin" and the absence of God is despair (151). Yet Sartre, a leading prophet of twentieth-century existentialism, has no religion; and to the question of how values can be invented without God, he answers: "You've got to take things as they are. Moreover, to say that we invent values means nothing else but this: life has no meaning *a priori.* Before you come alive, life is nothing; it's up to you to give it meaning, and value is nothing else but the meaning that you choose. In that way, you see, there is a possibility of creating a human community" (152). And between Kierkegaard and Sartre we find Albert Camus quoting Pascal ("A man does not show his greatness by being at one extremity, but rather by touching both at once.") and stressing a freedom that accommodates both faith and atheism. In Camus, paradox comes to flower when he writes, "at the very moment when the artist chooses to share the fate of all, he asserts the individual he is" (153).

Existentialists, like some political parties, strive to make room for a broad spectrum of views. However, common threads appear. Many existentialists have rebelled against existing conditions and emerged as strong critics of their own time. They observed the tragic core of existence, sensed the triumph of the absurd, and saw man's challenge as that of rising above tragedy and absurdity for life's sake.

Existentialism has noted with somber accuracy the pessimistic fate of man and advocated a humanistic assertion of will against such pessimism. Sarte contends that since man, alone in the universe, inescapably is his own lawmaker "condemned to be free" and to carry the weight of the world, "man will fulfill himself as man, not in turning toward himself, but in seeking outside of himself a goal which is just this liberation, just this particular fulfillment." Identifying existentialism with humanistic commitments, Sartre calls his philosophy "optimistic, a doctrine of action, and it is plain dishonesty for Christians to make no distinction between their own despair and ours and then to call us despairing" (154).

Existentialism has been described as disillusioned with progress and lacking in hope for mankind, yet it emphasizes positive humanism as the essential alternative to human despair. Existentialism raises the question of absurdity, in that man's "essence" (consciousness) is sentenced to "nothingness" (death). Partially to escape this absurdity and the absurdity of the outside world, internal feelings receive special attention. Existentialists, looking for refuge from the absurd, tend to disdain science, history, economics, and politics, and to find refuge in artistic and individual freedom. In *The Myth of Sisyphus,* Camus called "suicide" the fundamental question of philosophy, and man's chief problem that of learning to hope without resignation, and his goal how "to live and to create, in the very midst of the desert" (155). The hope Camus describes may be found by some in nations or other people, but in his view it is "awakened, revived, nourished by millions of solitary individuals whose deeds and works every day negate frontiers and the crudest implications of history. As a result, there shines forth fleetingly the ever-threatened truth that each and every man, on the foundation of his own sufferings and joys, builds for all" (156). We serve mankind by perfecting our own lives.

Some existentialists (e.g., Dostoyevsky) have been fearful and contemptuous of rationalism as well as the technology it fosters. They found sanctuary in mysticism. In *The Brothers Karamazov,* Dostoyevsky wrote: "Much on earth is hidden from us, but to make up for that we have been given a precious mystic sense of our living bond with the other world, with the higher heavenly world, and the roots of our thoughts and feelings are not here but in other worlds" (157). This phenomenal work was published in 1880, and Dostoyevsky's spokesman, Father Zossima, had many messages of an incredible timeliness for our own age, including this one: "For how can a man shake off his habits, what can become of him if he is in such a bondage to the habit of satisfying the innumerable desires he has created for himself? He is isolated, and what concern has he with the rest of humanity? They have succeeded in accumulating a greater mass of objects, but the joy in the world has grown less" (158).

It is not true, however, that existentialists in general succumb to hopelessness and despair. Hopelessness and despair more accurately are their point of departure in the direction of hope and victory over despair. For some, such as Roszak, hope is achieved by "dropping out," or through mysticism, or by promoting the instincts to "Colonel" and demoting the intellect to "Private."

What values are expounded by existentialism? According to Sidney Finkelstein, existentialist morality assumes three main directions (159):

1. A personal relation to God (or nature). All human attempts to improve man's lot are abandoned. The morality of total faith is espoused (Kierkegaard).
2. A primitivist morality concentrates on the immediacy of "being." It calls for liberation of the instincts and the primal forces of life, to save them from the corruption of logic, science, or the plight of others (Heidegger).
3. The individual is asked to make a commitment, but the commitment is to himself, not to society (Camus). To preserve his own integrity, the individual must reject conventional moral precepts as hypocritical, and rely on his private judgment. He thus becomes a moral rebel.

Number 1 allies itself closely with ideological inertia. Number 2 became the morality of Fascism, the process of "Thinking with the Blood." Heidegger, the apostle of morality number 2 and a leading existentialist, praised Fascism and was part of Hitler's establishment.

The diversity of existentialism is seen in the fact that Camus, the exponent of morality number 3, was a leader in the French Underground throughout the war and continued, together with Sartre, to oppose Fascism to the end of his life. Camus' morality of individualism was used by him to condemn governmental and social institutions and practices he vehemently opposed, such as capital punishment. The drawback of morality number 3 is its essentially anarchistic nature, with responsibility entirely dependent on individual whim. If the individual happens to be Camus with an intensely active social conscience, morality number 3 can have positive impact with telling criticisms of modern society. But trusting all others to behave with similarly admirable responsibility would in some cases resemble an effort to reform a thief by giving him the combination to the safe.

Existentialist morality focuses almost exclusively on the individual and his inescapable aloneness in the universe. The individual's duty, therefore, is to himself, and he must answer only to himself. Thus existentialism accepts no active responsibility for society as a whole and does not concern itself directly with social improvement, except, as Camus says, through individual assertion. The lack of social values has been the most conspicuous weakness of existentialism, and tends to invalidate the philosophy except for artists, writers, followers of esoteric movements such as Zen Buddhism, and others with eccentric professions or appetites. An intriguing philosophy with curious and interesting ideas, existentialism has attracted many of the world's finest writers. Both Camus and Sartre won the Nobel Prize, though Sartre turned it down; and Dostoyevsky, of course, was one of the supreme masters of world letters. Nevertheless, existentialism has not qualified as a definitive philosophical guide.

Dialectical Materialism

It has frequently been suggested that the principal task of philosophers consists less in providing answers than in asking the right questions. This saying eloquently applies to the philosophical work of Karl Marx who became skillful at posing fundamental social and economic questions, but whose answers never quite functioned as he wished. Both the questions and the answers have made the name of Marx controversial around the world; praised by many, damned by others.

Marx was a social agitator and economic philosopher who was busy in the middle and later years of the nineteenth century fighting with his pen for liberal and social causes. Marx became the leading theoretician of socialism. With his friend and lifelong associate Friedrich Engels, he compiled and published the earth-rocking *Communist Manifesto* in 1848.

Exiled from Germany, Marx in 1850 established himself in London, where he remained for the rest of his life. For the most part he busied himself in the world famous library of the ever-accommodating British Museum, where he did the research for and largely wrote his classic *Das Kapital,* one of the world's best known, best loved, best hated, and least understood books. Marx died (in 1883) before the three volumes of *Das Kapital* were finished, and Engels carried the work through to completion.

Marx and Engels sought a way to provide a suitable philosophical base for their devoutly held theories of socialism. They used the popular but unscientific approach (though many scientists have also used it) of starting with certain conclusions and looking for a philosophy to sustain the conclusions.

Dialectical materialism is the result. As we have seen, idealists envision a fixed system with immutable ethics, while pragmatists talk about a flexible universe with ethics adaptable to the need and the good of the moment. The dialectical philosophy improvised by Marx at the British Museum holds that permanence and change are opposites but that they do not clash. An interplay between the two through the social interactions of men brings new developments into being. Interaction between permanence and change results inevitably in a pattern of change, but the final pattern owes its shape to the counteracting impulses of both permanence and change. Analogously, in an organization, the members through interaction change their own relationships and in so doing they change the organization. Change, in other words, is inevitable wherever an interaction occurs, and it follows a predictable channel, though the results are seldom predictable.

For example, in connection with the familiar free will versus determinism issue, dialectical materialism would hold that man modifies his environment and thus to that extent is free. However, at the same time, the environment modifies man, and in this interaction, new relationships arise. The end result is that man is not completely free, nor is he entirely subject to the dictates of determinism. He is somewhere in-between, nostalgic for certainty.

This duality and its necessary interplay can be likened to a medical mixture composed of two separate ingredients. To make the medicine effective, you must "shake well before using" to assure a thorough mixing of the ingredients.

Marx developed the philosophy of dialectical materialism in a sense by taking separate ingredients and shaking them well together. In writing *Das Kapital* Marx used the dialectic method of Hegel to establish his points about mankind's history of class struggles and the final triumph of the proletariat in a classless society. Marx's application of Hegel's dialectic method was almost directly opposite to that intended by Hegel. As Hegel used it, dialectic referred to a logical progression from thesis to antithesis to synthesis. In dialectical materialism, the progression reverses. It starts with the theory that the pattern of human life is established by economic need (synthesis) which results in classes (antithesis). The struggles among classes constitute the historical process, and the end result (thesis) comes with the withering away of states and classes as the proletariat assumes power.

The emotional and political appeal of these arguments proved much stronger than their philosophical validity. Marx made a number of economic assumptions, extrapolated them into social terms, and built an elaborate house of unproved and largely unprovable theories. Nevertheless, the appeal made the philosophy one of the most influential in the history of man.

Dialectical materialism holds that change is experienced socially in the communities of men solely through materialistic economic forces. The term "materialism" signifies that human history, social relations, economics, and existence itself result from material influences. Materialism contrasts in a fundamental way with idealism. Materialism contends that consciousness is a reflection of material things, while idealism sees the social structure and all material items as a reflection of consciousness. It is a "which came first, chicken or egg" sort of question, with one side saying chicken, the other side insisting it has to be egg.

Materialism rejects the Absolute and has no space, even in the stable, for the supernatural. From this root came Marx's famous dictum: "Die Religion . . . ist das Opium des Volkes." (Religion is the opium of the masses.)

Marx founded his theories on the inevitability of class struggle. The ruling class exploited the laboring class, which in turn tried to seize a greater share of its own produce. These theories and the philosophy on which they floated served as suitable vessels for socialism and later, communism. The socialist slogan, "the development of human powers is an end in itself," was idealistically elaborated by communism into an economic dogma: "Jeder nach seinen Fähigkeiten, jedem nach seinen Bedürfnissen." (From each according to his abilities, to each according to his needs.) As with many slogans and dogmas, these have been more honored in the breach than in the observance.

The concept of freedom as originally stated by Hegel was, "Freedom is the appreciation of necessity." Freedom rests in man's ability to control his life and fulfill his needs. For this reason, socialism has continually stressed production, since only by meeting man's material needs fully would he be liberated for optimum development of his human potential.

The start of socialism in Russia stirred much interest and hope. Since that promising commencement, however, much of the hope has been frustrated. The Soviet Union has followed a course similar to that of most other national states. This course involves building military strength, amorality in international affairs, immorality in the name of national security, and suppression of internal dissent. Soviet leaders defend this course in the usual fashion: "Russia has dangerous enemies."

In the 1930s, youth rallied in the midst of a worldwide depression to the optimistic ideas and hopes of socialism. Later generations in the developed countries became disillusioned with the unkept promises of socialism, and turned away in favor of philosophical retreat, negativism, immediate pleasure, and other private concerns. Marx predicted that socialism would spread first and most successfully in capitalistic industrialized countries such as those in Western Europe, but its record of success in those countries has been inconclusive and erratic, sometimes for other reasons, economic and political, than inherent flaws in the political philosophy. In fact, socialism's appeal has tended to be stronger in underdeveloped countries, where it still attracts many followers.

Increased production stands as a pressing obligation under socialism. This is admitted as the only way to provide a base for the classless society. The Soviet Union, still the leading socialist example, heavily favors science and supports engineering to lift itself by its technological bootstraps and industrialize rapidly. The Russians are ambitious eventually to overtake the U.S. in industrial production, and for this reason, it should not be surprising that the U.S. and Russia have come to resemble each other in many ways.

Critics of technology playing the game of fair's fair, have been just as vitriolic toward socialist as toward capitalist countries. In the case of Russia and the U.S., similar problems of pollution, waste, alienation, and corruption have been experienced in both. The intense and frightened competition between the two now threatens humanity with potential destruction. Thus, both critics and defenders of technology have become intent U.S. and Russia watchers. When will the conflict happen? Can it be kept "safe," that is, waged with conventional weapons? Can we avoid the conflicts? How?

Putting ideological concerns aside, technologists in their work have rather consistently adopted materialism instead of idealism for philosophical support. Materialism encourages observation and experiment to explore nature's materialistic fabric. It happens, of course, that technologists are sometimes materialistic in their work, and idealistic or mystical in other spheres of their lives.

Critics of technology, as we might expect, also quite often oppose materialism as a valid philosophical base. Perhaps the modern engineer who spends his day verifying material phenomena and his evenings practicing Transcendental Meditation or finding spiritual peace in Zen Buddhism or studying world religions in pursuit of the supernatural, tries to accomplish a coming together of the chicken and the egg, a cohesion of materialism and idealism, and interaction between permanence and change in good Marxist fashion.

Certainly for the sake of man's future, we need all the cooperation we can get. Cooperation between philosophies—or at least understanding—is a useful place to start. If internecine conflict can be avoided between different methods of looking at man and the universe, a better chance exists that ultimate conflict can be avoided between competing national states. It doesn't hurt to try, and to paraphrase Karl Marx, we have nothing to lose but our man-made opportunity for early extinction.

UNIVERSAL RULES FOR MAN

The establishment of moral laws has long been a dominant human concern. There is perhaps no more popular avocation than the effort of individuals to codify moral precepts for the guidance of their neighbors.

Man seemed to want absolute moral laws from the beginning of ethical introspection. We find an individual awareness in man of certain rights and wrongs as personal concepts, yet difficulties confront us in stating definitive moral laws satisfactory for all and acceptable to all. Moral laws adequate for a modern city will not satisfy a Stone Age tribe in the jungles of New Guinea. Moral laws prove inextricably subject to the needs of local conditions. Unless the laws reduce to a few simple human basics, the instinctive do's and don'ts of the human race, moral codes often prove impossible to export successfully and to implement in different environments and cultures.

This is not to say that people everywhere are not essentially alike, and that the human basics don't apply. Margaret Mead, the anthropologist, has told of how Stone Age tribes have leaped 10,000 years to the modern age in a single generation. She describes how one young man told her during a field trip, "My father is a cannibal. I'm going to be a doctor." And according to Margaret Mead, he became one.

The point perhaps is that technology and ambition for modern progress are easier to export than western ideas of good and bad, which may seem pathetic nonsense to others. History provides innumerable examples of desperate conflicts resulting from a confrontation of different moralities: the brutal meeting of Indians and Europeans in North America, or that of Christian missionaries and Polynesians in the Pacific islands.

Despite these tragic encounters, the effort to discover and formulate universal moral rules has continued from early times to the present. In western culture, the best known and in terms of endurance, the most successful moral code has been the "Golden Rule" of Jesus Christ, variously paraphrased from the 12th verse, 7th chapter of St. Matthew: "Therefore all things whatsoever ye would that men should do to you, do ye even so to them: for this is the law and the prophets."

Other religions have offered variations on this precept in different words but with the meaning substantially the same. Endeavoring to fit the Golden Rule and similar moral guides into the formal corpus of philosophical thought, German philosopher Immanuel Kant introduced the "categorical imperative," which describes rules that can be considered ethically mandatory in human conduct. "Act only on that maxim whereby thou canst at the same time will that it should become a universal law," wrote Kant of the categorical imperative. Much better than most men, philosopher Kant obeyed his own dicta, remaining a bachelor and living out the 80 years of his life in his native town of Königsberg, Germany. Kant came to epitomize the absent-minded philosopher, wandering about town with his head in the higher stratospheres of philosophy; and so regular were his wanderings, his fellow townspeople were said to set their clocks by his movements.

The dilemma, of course, with universal moral rules lies in the fact that they emphasize what we should be doing rather than what we are actually doing. It is neither an accident nor a surprise that many consider certain moral advocates rather vague, impractical creatures as in the case of Kant or that they are crucified as in the case of Jesus. It isn't so much that people dislike being told what they should do in a moral sense. It is rather that they have a natural distaste for realizing that they aren't doing it, and they feel the tiny, but insistent needles of man's elusive inner critic, conscience. That low, strange voice saying, "don't" beforehand and "you shouldn't have" afterwards, encourages the hope that the necessity for moral laws may become more appreciated, and their application more prevalent.

Obviously a considerable amount of individual morality abounds among men. Good Samaritans may not appear as often as we should ideally like, but they do appear. Some men and women become physicians, not to get rich, but to heal the sick. Others invest their lives in the education of the young. And many others say "good morning" with a smile and mean it.

There are grim exceptions, of course, and it would be irresponsible to ignore them. Science, if it does nothing else, must insist on the truth. One truth is this: if morality abounds, so does something that can only be called monstrosity appear with flagrant frequency among men. The insane murders of Nazi Germany are still part of living memory. And *The New York Times* for August 22, 1976 reported incidents in Hartford, Connecticut that seem to emphasize the mind of man has not emerged from whatever jungle in which it began. Or did man's mind start from a garden in paradise and from that innocence now works its way toward some future jungle? A plausible moral case can be made for either view.

This is what occurred in Hartford. A young woman, crazed perhaps by drugs, slashed her wrists and ran to the steps of a Catholic Church. While police and priests tried to persuade her to hand over the razor, with which she threatened to cut her throat, a jeering crowd of people assembled, threw bottles at the distraught woman, and urged her to go ahead and carry out her threat. What anger or sickness in the crowd brought on this display of cruelty? The scene from beginning to the final collapse of the girl due to loss of blood, with the crowd cheering, was the equivalent of an existentialist nightmare, and one begins to understand both the courage and the despair of Albert Camus when he wrote in 1943, with the Nazi war raging, "He who despairs of events is a coward, but he who has hope for the human lot is a fool" (160). Yet even enveloped by tragic absurdity, the philosopher in Camus knew the necessity of living as if one possessed hope, and as if there was a positive purpose in so living. "One's duty is to do what one knows to be fair and good. . . This does not mean that the absurd does not exist. It means that the absurd is *truly* without logic. This is why one cannot *truly* live on it" (161).

This suggests why the ultimate and most essential human value, echoed in every valid human moral philosophy, may simply be trust. We have to trust that the driver in the approaching automobile will behave rationally even if he subscribes body and soul to the irrational. Without such trust, civilization crumbles in a moment. Legislatures and rulers establish laws for the regulation and preservation of society, but what really preserves civilization is the social contract in effect among men for reasonable obedience of reasonable laws.

Law simply cannot manage without the indispensable moral cement of trust. Trust is the vital ethic absolutely necessary to regulate human relations. We have to trust that our neighbors will prove trustworthy and that they will trust us.

Consider a jet flight from anywhere to anywhere. How many others, largely strangers, must be trusted totally with our lives. Engineers, designers, assemblymen, mechanics, inspectors, pilots, crew, radio operators, meteorologists, baggage clerks, etc. We have to trust not only the many people serving and servicing our plane, but the thousands of people performing the same duties for other planes. The human value of total trust even involves trusting a life insurance company with an impromptu airport insurance policy, just in case.

For the most part, this trust occurs without question or thought. Faith in those thousands of our fellow men charged with transporting us safely is deeply ingrained. If it were not, how many would fly?

When bank depositors line up in panic to remove their funds, their normal trust has been replaced by sudden fear. Contracts provide a written record of a transaction, but they are essentially just memoranda between men of good will. Remove the good will, and in most cases, how valid, or more important, how valuable are the contracts?

Most of us realize the impossibility of living together with others if each considers only himself, his own desires, and behaves accordingly with no concern for others. Without trust and the prudent cooperation it encourages, the efficient machine of society quickly becomes costive with rust and comes creakingly to a halt. To achieve and hold the trust of others, men come to terms with basic ethics by recognizing the dangerous folly of selfishness, which results in ostracism, and by striving to make correct actions habitual. Even professional criminals can be trusted, perhaps more than others, to be rigorously scrupulous about fundamental social rules, such as crossing with the light. They strive constantly to avoid being conspicuous.

Human ethics, though not always exemplary, do seem workably trustworthy under normal conditions, largely because sane men see the good sense of behaving sanely. Organizations and nations cannot, however, be so routinely trusted. Organizations, especially those in pursuit of profits, often seem with increasing size, increasingly immune to basic moral considerations. Big corporations eventually reach the point where in effect their behavior suggests they are convinced sheer size justifies them in crossing against the light, perhaps because of the certainty that whatever hits them, the hitter will be hurt more than the hit.

As for nations, self-interest becomes the epitome of ethics, and national security becomes holy writ. To serve the former and protect the latter, the "anything goes" philosophy takes over totally to an extent that would compel Genghis Khan to watch with envy.

Certainly one of the paramount tasks for man now and in the future is to find ways of making individual trust a working reality in organizations and nations. This may be the most difficult task ever undertaken by the human race and undoubtedly the most important. One aspect will be educating businesses and governments (specifically, the "company men" that guide them) that selfish self-interest in many situations will prove to be dangerously irresponsible and ultimately destructive of self-interest. Another aspect will be educating them to comprehend what E. B. White was driving at when he wrote to a group of evangelically isolationist Congressmen during World War II, "Gentlemen, if you do not know that your country is now entangled beyond recall with the rest of the world, what *do* you know?" (162).

Technologists know the truth of this entanglement. They know as well the truth about man's dwindling resources, the truth about the pollution and population crises, the truth about world hunger. Technologists also have a language that is understood by their counterparts in every nation on earth. Those are reasons why scientists and engineers must be involved in the worldwide quest for social morality. When should that quest begin? Since it didn't begin yesterday in earnest, it should begin today. Tomorrow may be too late.

REFERENCES

131. Freud, Sigmund, *Civilization and its Discontents,* W. W. Norton, New York, 1962, p. 11.

132. Highet, Gilbert, *Man's Unconquerable Mind,* Columbia University Press, New York, 1970, p. 9.

133. *Daedalus,* Summer, 1974, #3, pp. 33-45.

134. *Ibid.*, pp. 17-32.

135. Dunham, Barrows, *Giant In Chains,* Little, Brown & Company, Boston, 1953, p. 119.

136. Ardrey, Robert, *African Genesis,* Atheneum, New York, 1961.

137. Ardrey, Robert, *The Territorial Imperative,* Delta Book, Dell Publishing Company, New York, 1966.

138. Lorenz, Konrad, *On Aggression,* Harcourt, Brace, and World, New York, 1966.

139. Montagu, Ashley, Ed., *Man and Aggression*, Oxford University Press, New York, 1968.

140. Köhler, Wolfgang, *The Mentality of Apes*, Penguin Books, Harmondsworth, Middlesex, England, 1957.

141. *Ibid.*, p. 248.

142. "The Nature and Nurture of Behavior," from *Scientific American*, W. H. Freeman & Company, San Francisco, 1973.

143. Bodmer, W. F., and Cavalii-Sforza, L. L., "Intelligence and Race," from *Scientific American*, W. H. Freeman & Company, San Francisco, 1973.

144. Köhler, Wolfgang, *The Mentality of Apes*, p. 252.

145. James, William, *A Pluralistic Universe*, Longmans, Green, New York, 1909.

146. Mill, John Stuart, *Utilitarianism*, Bobbs-Merrill Co. Inc., New York, 1957.

147. Meador, Roy, *Franklin—Revolutionary Scientist*, Ann Arbor Science Publishers, Ann Arbor, Michigan, 1975, p. 108.

148. Mehta, Ved, *Fly and the Fly-Bottle*, Penguin Books, Harmondsworth, Middlesex, England, 1965, pp. 47, 49.

149. *Ibid.*, p. 34.

150. Dunham, Barrows, *Giant in Chains*.

151. Kierkegaard, Sören, *The Sickness Unto Death*, Doubleday Anchor Books, Garden City, New York, 1954.

152. Sartre, Jean-Paul, *Existentialism and Human Emotions*, The Wisdom Library, 1957, p. 49.

153. Camus, Albert, *Resistance, Rebellion, and Death*, Alfred A. Knopf, New York, 1961, p. 266.

154. Sartre, Jean-Paul, *Existentialism and Human Emotions*, p. 51.

155. Camus, Albert, *The Myth of Sisyphus and Other Essays*, Vintage Books, New York, 1960.

156. Camus, Albert, *Resistance, Rebellion, and Death*, p. 272.

157. Dostoyevsky, Fyodor, *The Brothers Karamazov*, The Modern Library, New York, 1950, pp. 384-385.

158. *Ibid.*, p. 377.

159. Finkelstein, Sidney, *Existentialism and Alienation in American Literature,* International Publishers, New York, 1965.

160. Camus, Albert, *Notebooks 1942-1951,* Alfred A. Knopf, New York, 1965, p. 80.

161. *Ibid.*, p. 83.

162. White, E. B., *The Wild Flag,* Houghton Mifflin Company, Boston, 1946, p. 19.

QUESTIONS FOR REFLECTION

1. Do you consider man inherently aggressive? Give examples and counter-examples.

2. Consider the technological, political, social, and moral implications of Michel de Montaigne's observation that "nothing is so firmly believed as what we least know."

3. What is progress? Has the human race progressed or retrogressed ethically in your lifetime? Give examples.

4. What are the five basic moral rules you try to obey in your own life? How are you doing?

5. Explain, in any language you prefer, your philosophy of life.

6. Would you do research on projects to build deadlier weapons? Would you activate and explode an atomic bomb carrying rocket if so ordered?

7. Can it ever be patriotic to disobey orders in time of war? Were the Vietnam draft evaders justified in obeying what they consider a higher moral commitment than patriotism?

8. For a shipwreck companion on an uninhabited island would you choose a cook, a philosopher, or a philosopher who can cook?

9. The French World War I battlefield practice of *triage* has been suggested as a potential international policy essential for the survival of civilization. (*Triage* involves allowing some to die and treating only those who show the best promise of surviving if treated.) What are the moral implications? Should technologists collaborate in such programs?

10. Are man's emotions more important than his intellect? Are they more or less dangerous? Give examples.

11. Conclusion: Beyond the Collision

> "When I sitting heard the astronomer where he lectured with much applause in the lecture-room,
> How soon unaccountable I became tired and sick,
> Till rising and gliding out I wander'd off by myself,
> In the mystical moist night-air, and from time to time,
> Look'd up in perfect silence at the stars."
>
> Walt Whitman

> "O well-beloved stone-cutters, I lead them who plan with decision and science,
> Lead the present with friendly hand toward the future."
>
> Walt Whitman

HOW FAR HAVE WE GONE?

We began by referring to the schism between the cultures, occupied by doers on one side and contemplators on the other. The argument was offered that the world has become too dangerous in the twentieth century for continued separation between doers and contemplators. Yet that is where we are, separated, with each side imperatively needing to know and understand the other.

Specifically, technologists can indulge no longer in the assumption that the other culture is occupied entirely by impractical idealists. Those in the other culture need to accept that technology has shaped the modern world and will continue to shape the future. The opinion was offered that if technology is sometimes used for bad ends, all bear responsibility, because the fate of all is at stake. No one is a noncombatant. All must participate in the solution when there is a collision between technology and human values.

In his *Notebooks,* Camus records a statement by Arthur Koestler that seems relevant for everyone. Koestler was talking to writers, but it might pay for even nonwriters to listen: "We are guilty of treason in the eyes of history if we do not denounce what deserves to be denounced. The conspiracy of silence is our condemnation in the eyes of those who come after us" (163).

Energy? Environmental pollution? Jeopardizing earth's ozone layer? Recombinant DNA research? Plutonium disposal? Nuclear reprocessing? These and other technological choices hang over our heads like the fabled sword Dionysius suspended from a hair above the head of Damocles so he would understand what it means to enjoy the happiness of a king. Koestler's warning could be interpreted to mean we are guilty if we keep silent, if we settle for a smug, right-now happiness and ignore the future. Applied to non-technologists, the warning sounds an obligation to speak up when technology leads or pushes us in directions we have reason to consider presently or potentially undesirable. Applied to technologists, the obligation entails not sitting idly by when scientific work is being used dangerously or recklessly.

In 1962, C. P. Snow noted that a few men in secret make critical choices concerning highly sensitive scientific matters such as nuclear fission, with dangerously limited understanding of science or the ultimate consequences of their decisions. "This phenomenon of the modern world is, as I say, bizarre," wrote Snow, "We have got used to it, just as we have got used to so many results of the lack of communication between scientists and nonscientists, or of the increasing difficulty of the languages of science itself. Yet I think the phenomenon is worth examining. A good deal of the future may spring from it" (164).

Only limited progress has been made since Snow published *Science and Government.* The two sides of the campus are still separated. The phenomenon Snow called bizarre continues. The communications gap persists, and crucial scientific choices are being made by politicians and others with conspicuously meager scientific credentials. Scientists and engineers have seldom been asked to make the major technological decisions, and in most of those decisions they have been consulted only superficially.

This is not to declare scientists and engineers innocent. Technologists can be criticized for what they have not done with sufficient zeal—attempted to make technology understandable to others. This fault lies not so much in commission as in omission. Technologists have tended to be reclusive in their work, and amoral with respect to the use of their products.

It is true that many serious questions confronting modern man are social and political rather than technological—racial divisions, economic disparity, injustice, totalitarianism—but such questions are never resolved today without the collaboration of technology, normally on a vast scale. Hunger may come about through social, political, and climatic influences, but it is never alleviated in a significant way without efficient technology. Yet dependent as humanity is on technology, most still know pathetically little about it, as well as the costs and the consequences.

Non-technologists usually are unable to take the lead in explaining technical matters to the general public. A few with special skills as thinkers and writers have managed the feat. Joseph Wood Krutch did so in his books on nature and the Arizona desert. Arthur Koestler, another twentieth century humanist-philosopher, wrote with great effectiveness on technology. The title given a collection of tributes addressed to Koestler at the age of 70 conveys a sense of the influence he has had in this century: *Astride the Two Cultures* (165). But for the most part, explaining technology is largely the job of technologists.

THE SOCIAL ETHICS OF TECHNOLOGY

Gilbert Highet in an essay entitled "Science for the Unscientific," commented on the "interstices of ignorance" with gaps in need of bridging. He wrote:

> One of the largest of these gaps is surely Science. The scientists have lots of books, of course, within their own world; but they do not often write books for the unscientific, and when they do they sometimes fail to make their subject pleasing or interesting. Someone once told me that Paul de Kruif was disdained by pure scientists because he had written successful books of what is called "vulgarization." If so, they are unwise. He has done them a great service. (Here I speak with a certain emphasis, for my own subject, the classics, was very nearly killed by specialists who despised the public and would not explain their own work.) (166).

Nearly a quarter of a century has passed since Highet wrote, but his message remains pertinent. So is the answer given by Senator Margaret Chase Smith to a reporter who asked what subjects children should study in this complex world: "First and foremost science. There should be Science Appreciation courses taught in the elementary grades to get our young people interested in science and to make them realize what science does for their lives, and to let them know what science can do" (167).

Comments such as those by Highet and Senator Smith indicate that stronger communication efforts on the part of technologists would be useful. In addition, technologists need to wage a campaign for recognition that the international demand for the benefits of technology is not an accident, and that the benefits are genuine. Technology for some has assumed an insensitive or even sinister aspect because of negligent communications. This is bad for technology, and bad for the misled. The social ethics of engineers and scientists clearly are comparable to and in some cases perhaps superior to the social ethics of other professions. Consider physicians. They encourage publicity about service to society, but their image also includes a pronounced stress on high income as indicated by the tendency of doctors to concentrate in wealthy centers, like bees around their queen. An emphasis on convenience is also not unassociated with the medical profession, as satirized in a *New Yorker* cartoon—the complacent doctor explains on the telephone that "I do not make house calls, Madam. On the other hand, I've never been accused of providing unnecessary medical attention."

Doctors, of course, are simply a conspicuous and recognizable example. Practitioners of other professions from plumbing to lawyering are similarly inclined to dismiss Camus's warning that "any life directed toward money is a death" (168).

Technologists in fact seem to have a stronger record of absorption in their work, with less concern for what they can get out of it. This does not call for a session of back-patting. In their academic and laboratory environment, technologists may be less susceptible to the virus of economic temptation, and their opportunities may simply be fewer. The degree of social ethics in other groups has no particular bearing on the responsibilities of technologists. Their concern is with their own profession, and the values they serve.

DO THE LIGHTS LIGHT?

Examining their work, some of those values can be identified with relative ease. Integrity is one. In a society dependent on technology, reliability is an absolute. The wings of the airplane have to stay on. The automobile has to stop when braked. With few exceptions they do, attesting to the responsibility with which technological tasks have been accomplished. The dependability of technology is sufficiently routine to be taken for granted. Buildings seldom fall down. When we flip the switch, electrical servants normally go to work.

Scientific integrity stands out as a fundamental ingredient in the makeup of most scientists and engineers. A few have falsified work to gain reputation or reward, but such efforts nearly always fail. The false stands out with unmistakable clarity, like the rattle of garbage cans in the midst of a symphony. Scientific work cannot occur effectively without persistent questioning and a healthy skepticism. Each theory and result must survive rigorous inquiry. Each product must endure hard use. A Nobel winner and world authority can state what should happen, but if it doesn't happen, authority once again is humbled. And the oldest commandment of technology goes immediately into effect—"back to the drawing board."

Will the lights go on? They will or they won't, it can't be faked. For this reason, the technological professions tend to attract those who are impressed by results and making things work, rather than by personalities. Engineers, for example, rarely shield incompetence, which has sometimes been done by other professionals "for the sake of the profession." Only true results matter in technology, and there is no way to shield incompetence for long. Sooner or later, the lights won't go on.

Social ethics, of course, inevitably demand more from technologists than integrity and dependable performance (each is a *sine qua non*). With technology, a multitude of problems entered the world. Non-technologists have tried to find solutions, but with infrequent success. Now technologists must join the search. It is also clear that some of these problems are critical, and that a planet lethally inhospitable to human life can, given sufficient neglect, be the last of man's works.

Expressing that threat with relentless force has become a task for united mankind, though mankind obviously is not united. The task, looking for a doorstep on which to rest, seems to settle instinctively at the door of technology. Much comment has been made concerning the international brotherhood of science. Such commentary will need to graduate perhaps from lip service to reality. The effectiveness of scientists in helping solve human social problems will be greatest if scientists organize and work together.

Engineers and scientists today do much of their work in groups or teams. The isolated genius in his ivory tower on a private mountain is an endangered species. Establishing strong, effective groups has been identified as a priority if technologists are to assemble the requisite "clout" to affect decisions and values. Groups of concerned scientists have been formed in connection with various issues, such as nuclear power. More will be needed. An individual is less effective when he represents just himself, than when he represents himself and the American Association for the Advancement of Science or other professional organizations.

Scientists may have done more in this respect than engineers. Engineers have shown slight zeal for expressing opinions on matters outside their professional interests. For this reason, and since many engineers have never taken the trouble to inform themselves in depth about the major human issues, they have largely been without influence on questions involving society at large. The condition is not incurable.

HALTING THE DRIFT

To change such attitudes, effective engineering organizations have been advocated that start with a commitment to be heard concerning the future directions and choices of society. Another useful step would be to relate the importance of social considerations to the design problems engineering students consider. In many schools, technology undergraduates take "socio-humanities" courses because the curriculum requires it. But students often find little relevance in such programs, and become more inclined than ever to avoid the "other side of the campus."

If design problems emphasized the effect of such designs on the lives of people in their communities, and the ultimate effect on the nation and the world, the student might take greater awareness with him into his career. Students will respond eagerly to concrete, realistic programs in the humanities that are convincingly offered and even more eagerly when the material is plausibly connected to their professional studies.

Design classes seem a logical place to begin. If technological alternatives were analyzed in terms of their social consequences, and further related to feasibility and costs, the students would receive proof that they do more than simply technical exercises, that their work can have a beneficial, or a catastrophic impact on society.

As Snow and others have pointed out, the gap between the scientific and nonscientific cultures has become an obstacle to the solution of technological and social problems. If broader social elements were included in design classes, the students would benefit, and progress might be made in bridging the gap between doers and contemplators. In a civilization dependent on technology, it is important for everyone to know as much about it as possible. Just as important, technologists must know the social and political facts of the world they live in. Engineers and scientists have special responsibilities in a technological society because they know better than laymen the potential consequences of technological actions. They can judge a particular technological prescription in terms of resource depletion, atmospheric pollution, human health, and other variables. These are always part of the full price that must be paid for any new technology, and are often more important than the money cost.

Technologists, though somewhat fallen in public esteem during recent years, are still respected for their knowledge and technical expertness. This respect gives them special authority to suggest realistic programs and have them accepted. Organizational growth among technologists would enlarge and strengthen this authority. When a group of scientists declare through a spokesman that a certain action is essential, whether it means inoculation to avoid an epidemic or leaving town because an earthquake is coming, most citizens usually act. They assume that scientists know what they are talking about, and they trust scientific knowledge.

This trust will be needed when the technologically aware increasingly speak out on the great issues and dilemmas of the future. As more scientific voices join those already issuing uncensored statements concerning real problems and tough solutions, mankind's slow drifting toward irreversible disasters may be stopped in time. Most people, with no acceptable solutions in sight, go on with their lives, ignoring all the threats they can. Hunger, pollution, vanishing resources, war—each produces a challenge most feel helpless to prevent or reduce. This is not likely to be the attitude that technologists will find suitable for themselves as future threats come closer to the present. Scientists and engineers will find it necessary to play a role in stopping the drift, and they will communicate the basic lesson of science—that there are no problems without solutions if the human race can learn and accept them.

The future will require technologists to do what they were not invited or allowed to do before, assume the leading part in resolving technological problems. Don't be surprised if they express views on other problems too.

IS TECHNOLOGY A "GOOD THING"?

Considering the warnings of technology's critics, should man take a chance on technology? Should he try openly to reshape nature and himself? Should he follow the *Rubaiyat's* suggestion to "grasp this sorry scheme of things entire" and after shaking it apart "remold it nearer to the heart's desire?" Should he stop tinkering with nature and try harder to accept nature's plan, whatever that might be?

The remarkable success of technology in shaking apart the old and remaking the new gives such questions more importance than they would otherwise have. Fearful critics, though wavering in their convictions and sometimes speaking with hazy or confused voices, seem to be saying of technological pursuits, "be careful, go slow." Other critics, more direct, issue a stronger message: "Stop."

Technology has shown no disposition to stop, and there seems little chance that technological pursuits, even the potentially dangerous ones, will be voluntarily dropped. The concern of critics is understandable, since technology often appears to demonstrate an accelerating momentum. Simultaneously, the awesome capacities of technology and scientific research lead inexorably to a time when "going too far" may be all too easy.

In 1976, for instance, research programs were underway in basic genetics that could alter existing life forms and create new ones. The researchers weren't certain what the full consequences of recombinant DNA research might be.

Lord Eric Ashby, speaking at the University of Michigan in the summer of 1976 on "Social Values and the Direction of Scientific Research," sought to reassure critics by describing the British approach to recombinant DNA research. Ashby emphasized that in Britain public participation in decision-making is the trend, even in complicated matters, and as chairman of the committee that had debated recombinant DNA research, he had taken unusual care to make certain the committee's report would be understandable. "I submitted the draft to an intelligent old lady of 80 who had no science training, with the request that she read it and mark with a red pencil every sentence she did not understand on first reading. Then I rewrote it to answer her criticisms," said Ashby (169).

The controversy surrounding the question of recombinant DNA research derives from the fact that genetic materials of different organisms are combined, with results unpredictable. The British committee report held that the dangers involved are no greater than those of bacteriological warfare research which has continued for years, with people "handling absolutely horrifying things" according to Ashby.

GENES AND ATOMS

In the summer of 1976, a report was also issued that a research team at the Massachusetts Institute of Technology after nine years of effort had synthesized a gene, the fundamental unit of heredity for living things. Led by Dr. Har Gobind Khorana, the team "constructed the gene by assembling the four basic molecular units of the genetic code into the sequence, deduced by others, of one particular natural bacterial gene. Though the gene was synthesized entirely from off-the-shelf chemicals, it functioned naturally when implanted in a bacterial cell" (170).

Creating a gene and implanting it in a bacterium which functioned normally was a milestone in genetic research. Praised as "a pretty piece of biochemistry" by Dr. James Watson of "double helix" fame, many scientists saw the work leading to new breakthroughs in genetics (171).

The news report detailing the accomplishments of Dr. Khorana and his associates, reiterated the unpredictable consequences of recombinant DNA work. The same page of *The New York Times* carried as well a brief account from the Associated Press indicating the efforts of an environmental coalition to shut down atomic stations in Pennsylvania and New Jersey during studies on the safe disposition of nuclear wastes. Technology was news on many fronts including the micro fronts.

Genes and atoms are too small to see, but not too small for technology to identify, study, work with, and alter—for good or ill. Mystics, followers of certain religions, and many who are simply frightened, express concern for mankind and argue that we tamper with such basics as genes and atoms at our peril. Some vaguely dread that the Creator will be offended and will explode us all in wrath. Others, equally pious, say the reverse, that man should go as far as he can, do as much as he can. To them, man's accomplishments are a tribute to whatever being or force created the universe. Perhaps they imagine the Creator returning after half a trillion years, looking around, and exclaiming proudly, "I do good work! That creature I made for fun in a few minutes of playing around has done wonders."

The chances, of course, are equally strong that such a Creator, unless the second visit comes soon, would find a graveyard earth littered with the debris of vanished life and nonbiodegradable soft drink cans.

HUMAN VALUES AND TECHNOLOGY

Technologists tend to leave the responsibility of welcoming visiting Creators to politicians and poets while they go on about their business. Some technologists would echo Vincent Van Gogh's thought that the creative force in the universe "must not be judged on the basis of this world; it's one of his rough sketches" (172).

Technologists are convinced, of course, that technology is a good thing, and they have reasons. They are certain man should use his talents and extend his inventiveness as far as it can safely go. The fact that man's inventiveness can go too far and perhaps hasten the hour of extinction is not lost on technologists today. Growing numbers of them admit that they and their colleagues have not consistently done all they should to assure that technology *is* and remains a good thing. This admission they consider the prelude to action.

The worldwide pursuit of technology indicates that a voluntary "back to nature" rush, for much longer than a weekend, is unlikely. Critics have been manifestly unsuccessful in reducing technology or inspiring much of a sustained surge to the wild woods.

Nevertheless, critics have served the useful purpose of forcing technologists to reexamine their values and to correct whatever shortcomings they can. Scientists and engineers, confident that technology will continue to go forward, are now more openly considering all the needs of men, mental and moral as well as physical. The fact that a British organizer of scientific effort such as Lord Eric Ashby could speak on social values in connection with research at an American university in 1976 and stress the need for public awareness is symptomatic of a new and encouraging trend in the technological professions. Technologists question human values in ways they rarely bothered about before. The very question "is technology a good thing" would have seemed irrelevant to most scientists and nonscientists a few years ago. Now both groups are increasingly committed to the view that human society must move on from the level of excessive consumption and self-indulgence, and that society must recognize higher future values than mere creature comfort. Especially in the technological societies, too many people were trapped by the conveniences and benefits of technology before the true costs and penalties were known. Objective reappraisals and adjustments of consumption to rational levels are more common now than before. The realities of the future will make them more common still.

Technologists are being pushed, the same as the rest of the human community, to face the priority obligation of implementing universal values suitable for the long haul. Traditionally this has been left to philosophers, religious authorities, cabdrivers, vagabonds, bartenders, and others with time for that sort of thinking. The traditional way has become obsolete. Values are needed now by everyone, and they need to work.

What values? The selection can be confusingly broad. There are the values that Norman Mailer was bluntly seeking to impale (and exaggerate) when he described the American corporate spirit as "that immeasurably self-satisfied public spirit whose natural impulse was to cheat on the environment and enrich the rich" (173). There are the values that Larry Darrell, protagonist of Somerset Maugham's novel *The Razor's Edge,* was emphasizing when he talked about criticism of western materialism, and when he defended his own country, America:

> They think that we with our countless inventions, with our factories and machines and all they produce, have sought happiness in material things, but that happiness rests not in them, but in spiritual things. And they think the way we have chosen leads to destruction. . . You Europeans know nothing about America. Because we amass large fortunes you think we care for nothing but money. We care nothing for it; the moment we have it we spend it, sometimes well and sometimes ill, but we spend it. Money is nothing to us; it's merely the symbol of success. We are the greatest idealists in the world; I happen to think that we've set our ideal on the wrong objects; I happen to think that the greatest ideal man can set before himself is self-perfection (174).

The human values men can agree on and rely on are actually not very complicated. Extraordinarily difficult to maintain, they are not difficult to define. Essentially they are variations on the familiar Jeffersonian themes of "life, liberty, and the pursuit of happiness." These themes serve to identify both human values and Utopian ideals. Happiness, like self-perfection, is an elusive ideal; the pursuit of happiness (or self-perfection) is a purposeful value. Another famous American historical passage, the Preamble to the Constitution, hopefully repeats ancient human values: ". . . form a more perfect *Union,* establish *Justice,* insure domestic *Tranquility,* provide for the common *Defense,* promote the general *Welfare,* and secure the Blessings of *Liberty.*" Each abstraction is an old dream of mankind.

Behind abstract values there must be institutions and programs dedicated to achieving and preserving them, or the values wither and drop like leaves from an unwatered bush. Technology makes an indispensable ally in the establishment and maintenance of human values. Inherent in technology is the impulse to get things done. There is no reason other than apathy or neglect why technology can't be programmed to serve useful human values in addition to putting satellites in space.

SPECIFICS FOR UNITED EFFORT

A. A. Berle, Jr. prepared a list (175) of basic human standards that include the following:

1. People are better off alive.
2. Health is preferable to sickness.
3. Literacy is better for people than illiteracy.
4. Good housing is better than poor housing.
5. A beautiful environment beats an ugly one.
6. Cultural opportunities bring people enrichment and enjoyment.
7. Universal education with each student allowed to go as high as he can manage is worthwhile.
8. Expanding the sciences and the arts is commendable.
9. Ample living resources must be denied no one.
10. Leisure and contact with nature are desirable human benefits for all.

Berle emphasized what can never be too often emphasized—the obvious. He did not limit the word "people" to a particular race or continent. Variations on his list could easily be made with different words, but many items would inevitably be repeated. Whatever list of value standards we use, certain clearcut needs stand out. To achieve such values, trust is essential and recognition that "there are more things in men to admire than to despise" (176). To maintain the life Berle stressed first in his list, a wide range of effective programs will be needed from the prevention of pollution to ending the menace of nuclear destruction.

To reach and uphold Berle's standards, or any comparable list, will call for much greater world cooperation, which could mean that countries, including the great powers, would voluntarily have to relinquish a portion of their traditional sovereignty. To promote health and reduce hunger, rich nations would have to contribute more to poor nations. World resources would have to be used with greater frugality and wisdom. Wastefulness, failure to recycle, profligate energy and material use would call for social ostracism, not acclaim. Population would have to be limited to levels the earth can humanely accommodate. To achieve the program suggested by Berle would demand the united effort of all if the values involved were to have wide rather than limited application.

At the moment, countries possessing large resource advantages, with the standards such endowments support, show little inclination to sacrifice on the scale needed for the benefit of those with less. Most countries remain largely absorbed in serving and protecting themselves. The U.S., for instance, has contributed substantially to other countries; but the U.S., as the wealthiest nation in human history, still has shared only a minute percentage of its wealth with others. Aid programs nearly always stop before reaching the level of substantial sacrifice. To the extent that true generosity has been involved, rather than political and national self-interest, praise is due the U.S. and other contributors to UNESCO, the World Health Organization, the World Bank, and related organizations with records of commendable achievement. Yet much more is needed from the U.S. and other economically favored nations in advancing humane interests that improve the broad, supranational welfare of mankind.

Hard questions persist. Will nationalisms accept voluntary limitations in the interest of humanity? Will population be controlled? Will the hungry be fed? Will technology be shared by mankind rather than hoarded for the benefit of a few? Will education make liberty an active cause rather than simply a word? Will the requisite united effort be forthcoming from earth's total human cargo?

At the moment there is little evidence that such a united effort is imminent. Yet few doubt that if men united, they could accomplish A. A. Berle's list of objectives, and more. Men have often proved their ability to unite on a large scale when confronted by a great and immediate challenge. In World War II, ideological opponents fought together against a common enemy in a common cause. International religious movements have proved superior to national borders in uniting men as their minds jointly reach out for answers to the ancient mysteries of the universe and life. When natural calamities have faced men, they have come together for united defense and mutual survival.

At Arecibo in Puerto Rico there is a large aluminum bowl on the earth's surface. The bowl is 300 feet deep and 1,000 feet across. Its metal surface is aimed upward at the sky. This remarkable construction, the world's largest radio telescope, is used by radio astronomers to send electromagnetic messages in binary code to distant corners of space thousands of light years from earth. The purpose of the messages is to inform anyone who might listen that they are not alone. Man reports himself present and accounted for. The Arecibo telescope is also used to analyze radiation reaching earth from space. The hope: that some day such radiation will prove to be decipherable code, indicating that we are not alone.

Suppose a message did reach mankind via the Arecibo telescope and that the message said: "We're coming. Will be there to take charge in 10,000 years. Prepare human race for consumption." Men would marvel, shudder, decide there was plenty of time to think of a solution later, and do nothing. Suppose the message said: "We're coming. Will be there to take charge three weeks from Thursday. Prepare human race for consumption." With the threat of extraterrestrial invaders and the "consumption" of the human race occurring in three weeks, unification of the world's peoples to combat the invaders would be managed in a fortnight. But given the same impending doom (in terms of resource depletion, atmospheric and surface pollution, starvation) with an indefinite time scale, men cling obliviously and stubbornly to the status quo.

The scientists at Arecibo have reported no messages from space. If men are to unite, and to use their brains and tools of science for broad human advancement, the incentive still must come from inner not outer space. It must come from recognition of common values and common challenges to human life.

United human effort would end the problem of collision between technology and human values. In place of collision could be progress based on technology that serves and is guided by human values. Some have labeled the alternative "unthinkable," and then have diligently *not* thought about it. Perhaps it is within the professional commitment of engineers and scientists both to think about and to do something about the "unthinkable." Though problems are formidable, we have powerful methods available for their solution. Hope rises like sap in the spring, and the words of an old hymn issue a fresh invitation:

> Creation's Lord we give thee thanks
> That this thy world is incomplete,
> That battle calls our marshaled ranks,
> That work awaits our hands and feet.

Awake or dormant, the values of man are present, and work awaits. That's the invitation. We wonder how the future will choose to answer.

REFERENCES

163. Camus, Albert, *Notebooks 1942-1951,* Alfred A. Knopf, New York, 1965, p. 146.

164. Snow, C. P., *Science and Government*, The New American Library, New York, 1962, p. 10.

165. Harris, Harold, Ed., *Astride the Two Cultures, Arthur Koestler at 70*, Random House, New York, 1975.

166. Highet, Gilbert, *People, Places, and Books*, Oxford University Press, New York, 1953, p. 169.

167. Smith, Margaret Chase, U.S. Senator, Interview, Folkways Record, FC7352.

168. Camus, Albert, *Notebooks 1942-1951*, p. 70.

169. Cummins, Robert, "DNA," *The Ann Arbor News*, September 21, 1976, p. 3.

170. Rensberger, Boyce, "Synthesis of Working Gene Hailed as a Major Advance," *The New York Times*, August 29, 1976, p. 1.

171. *Ibid.*, p. 45.

172. Camus, Albert, *Notebooks 1942-1951*, p. 76.

173. Mailer, Norman, "The Search for Carter," *The New York Times Magazine*, September 26, 1976, p. 21.

174. Maugham, W. Somerset, *The Razor's Edge*, Doubleday, Doran & Co., Inc., Garden City, N. Y., 1944, pp. 230-231.

175. Berle, Jr., A. A., "What GNP Doesn't Tell Us," *Saturday Review*, August 31, 1968, p. 10; also: Susskind, Charles, *Understanding Technology*, Johns Hopkins University Press, Baltimore, Maryland, 1973, p. 117.

176. Camus, Albert, *Notebooks 1942-1951*, p. 67.

QUESTIONS FOR REFLECTION

1. In order of priority, what are your personal values?

2. Which of your values are sufficiently important to justify killing or being killed in war before surrendering them?

3. What is the relevance for modern man of Patrick Henry's Prayer, "Give me liberty or give me death?"

4. "Environmental impact" studies are now required before approval of large new projects in many states. Should similar studies be required before any new technical development is released?

5. What grade might a visiting deity give the human race at this stage of evolution? If you were a visiting deity, how would you grade yourself?

6. What do human values prescribe in the contemporary choice between standpat safety and dangerous progress? Apply these considerations specifically to the drastic modern problems of environment, energy, and nuclear proliferation.

7. Using the radio telescope at Arecibo, what messages should we hope to receive from deep space? What messages should we seek to dispatch concerning man and his technology?

8. What myths concerning technology do human values expose and correct?

9. When technology and human values collide, how is a solution determined?

10. Which should have priority in today's world—technology or human values? Why? Which has priority?

Index